Julie Goldman

A Pilot's Life

Julie Goldman

A Pilot's Life

As told to Catherine Andrews

ORANGE FRAZER *BOOKS*
Wilmington, Ohio

ISBN: 1-882203-69-0

Orange Frazer Press, Inc.
Box 214
37½ West Main Street
Wilmington, Ohio 45177

Telephone 1.800.852.9332 for price and shipping information
Web Site: www.orangefrazer.com;
E-mail address: editor@orangefrazer.com

Jacket design by Tim Fauley

Library of Congress Cataloging-in-Publication Data

Goldman, Julie, 1912-
    Julie Goldman : a pilot's life / as told to Catherine Andrews.
        p. cm.
    Includes index.
    ISBN 1-882203-69-0
    1. Goldman, Julie, 1912- 2. Air pilots--United States--Biography. 3. Aeronautics--New
England--History. I. Andrews, Catherine, 1967- II. Title.

    TL540.G623 A3 2000
    629.13'092--dc21
    [B]

                                                                00-055737

Photograph opposite table of contents: Riverside Flying School instructors and members of the Massachusetts State Guard stand together at Muller Field in Revere before the war. Many of the men in this picture were killed in action.

Acknowledgements

Thanks to my late wife, Florence Goldman, who insisted that I write a book to accompany her story, *Pieces of Gold(man)*. To my daughter, Shelia Bauer, who started the endeavor. You listened and prodded me to tell hours and hours of the first stories of my life. You have made me proud by following in my footsteps, teaching others to love aviation as a career.

To my compiler, Cathy Andrews, who heard all the rest of my stories, organized my words and found the photographs to share. To Nadine Gibbs, who took current photographs. To Rebecca DuBey, indexer and knowledgeable aviation fan. To my twins, Marlene Triber and Myrna Giller; my two sons–in–law, Jeff Bauer and Lou Triber; my six grandchildren, Brian Triber and Nancy Triber-Hoar, Neal and Michael Giller, and Jami and Steffanie Goodman; and the Goldmans and Lermans for being a remarkable family to me.

To all of my aviation friends from Aero Club of New England and around the globe who continue to be my extended family and have given me much basis for my stories. Special thanks to my niece and publisher, Marcy Pliskin Hawley, who made this all possible. Thanks to the countless people who begged me throughout the years to tell my story. This book is for you.

Contents

Foreword

I was in between projects when my publisher handed me 10 hours of audio tape to transcribe. I was told the subject was 'Uncle Julie,' a rags to riches story. I sat at my desk at Orange Frazer Press listening and typing away. I heard the struggles of Uncle Julie, and his triumphs in overcoming the odds. I found myself laughing one minute, and crying the next. I became attached to a stranger with a voice from back home in New England. I rode on the rollercoaster of Julie's early life until the war started…and my tapes came to an end.

I wondered if he ever had a family with Florence. I wondered what he did after the war. I wondered how a kid left in Boston with $1 became a millionaire. When I moved home to New England, I was able to finish the book. I got to know Julie and his incredible story. I sat with him at his home in Peabody and learned about the rest of his life. I met his daughters. I researched the famous people he met and places he'd been. I pored through photo albums to find the best pictures to tell his story.

I learned that Julie has never taken the easy road. When presented with a challenge, he faced it. When given an opportunity, he embraced it. Through his 87 years, he has lived his life to the fullest. More than his strength of character, he has a good heart. I am fortunate that he was able to share it with me.

— *Catherine Andrews*
Epping, New Hampshire
May 2000

To the first
and second loves of my life,
Florence Goldman and aviation.

CHAPTER ONE

Early Days:

"**PA TOLD ME WHAT A** wonderful life it was in Russia. His family had a beautiful farm in Mezhirichi, Ukraine, with hay fields and cattle. My grandfather Moishe had a grist mill and an inn. He was 40. The inn was on top of the stable, and the family and guests stayed upstairs. My grandmother sold grain to earn money for food. The head priest in Mezhirichi, the gallach (Yiddish), was my grandfather's friend. In 1900 or 1901, he told my grandfather that the Cossacks would be coming through there. He said they would burn the place down and kill the entire family.

The Russian leader, Nicholas II, organized pogroms to slaughter Jews all across the country. The priest agreed to buy all the property so the entire family could go to the States. My grandfather sold everything to him, and that money got the family to Germany. From Germany, they only had enough money to get two of them to the United States. My grandfather's oldest son, Hyman, was 17. He went with him to Ellis Island.

They got to Boston and my Uncle Hyman set up a woman's dress shop. He was a talented tailor. He made enough money to bring the rest of the family to the States. He sent his father back to Germany to get the rest of the family. He and his wife, Gittel, 40, had three other children: Nathan, 15, Israel, 9, and Sarah, 11. They went from Germany to England. They left Liverpool

OPPOSITE:
JULIE'S MOTHER,
MOLLIE
REINHOLTZ,
POSES FOR A
PICTURE ON HER
WEDDING DAY.

and traveled by sailboat on the *S. S. Lake Superior*. They headed up to Quebec and sailed down the St. Lawrence River. They entered the Port of Quebec on July 9, 1901. The family went south to Massachusetts with my grandfather.

My dad, Nathan, was 15 and my Uncle Iz was 9. Iz was a good swimmer. He swam in the river near their home in Russia. On the trip from Germany he jumped off the sailboat into the water and swam in the ocean when it was calm. As soon as the wind picked up he climbed back on the boat.

From 1860 to 1930, 27 million people emigrated to America. Many were Catholics, many were Jews. They came from all over the world. They didn't have to get visas, they'd just go. There were so many Irishman, so many Jews—lots of Jews came to Chicago and New York. They spread out from there. Many went to Montreal and then traveled south. It all depended on where the ship was going. When my grandfather first got here, he was looking for the gold in the streets, and that's where he got the name Goldman. Our name wasn't Goldman. It was Gakalo. He asked where the gold was, 'Gelt, gelt.' The people here said, 'Your name is Goldman.'

MY MOTHER ALWAYS sheltered me from trouble. We lived in an old three-story building on Thornton Street in Revere, Massachusetts. The Revere Fire Department was across the street from us. When the firehouse bells clanged, the horses went charging down the street pulling the fire wagons. They had four horses that would pull an engine, called the Steamer.

Steam engines used to build up pressure for the water hoses. Smoke came out of a chimney on the wagon. The horses charged down the road with the engine behind them. It was 1916, I was four years old.

I used to run to see the horses. There were three flights of wooden stairs outside of the house we lived in. We were on the top floor. I went running to see the horses and I fell down all three flights. I was crying. My mother came charging down and picked me up. She hugged me and brought me upstairs. She said, 'Why did you do that?' I said, 'I wanted to see the horses coming out.' It was quite thrilling to see them charging out with the smoke and the clanging bells.

I was at Revere Beach with my father one day. He put me on his back and swam towards Holt's Pier, quite a distance away. Holt's Pier was on a curve of land heading towards Winthrop. A big flying boat used to drop passengers off at the pier. I never got a ride on it but I used to see it going across the water. Pa swam with me on his back when a Portuguese man-o-war stung him pretty bad. He turned around and swam back to shore with me hanging on. We ran for his old car, a Willy's Knight, and headed for home. I was taken care of in the house and he was brought to the hospital and treated.

In 1916, my sister Lillian was born and she was little. Just suddenly, one day there was a little sister. She was just another kid to me.

My family moved to Chelsea. I can see Pa standing with Ma. They had a little store and sold linens and stiff collars. The Strand Movie Theater was next door. My parents brought us to the theater in the afternoon while they ran their store. My older brother, Saul, and I sat in the comfortable chairs and watched the same movie two or three times. We'd come out or my parents would come and get us.

I saw mostly cowboy pictures, or serials. Somebody would drown or disappear and the next time they'd show up. William S. Hart was in the movies. He was a famous cowboy in those days, an old timer. Fatty Arbuckle was in the silent films. I couldn't read but I listened to the noise and watched the scenes of this big fat guy. We sat in chairs, but I don't think we ever had popcorn. We sat there until the lights went on and then we'd see the same movie all over again.

Sometimes they had an older girl take us around. I was in a playground with the girl and some kid threw a stone at me. It hit me in the head and knocked me out. The girl brought me home which was close by. We lived on the first floor of an apartment building. She took care of me until my mother came home.

We were very sick with the flu in Chelsea—me and my brother and little sister. My mother had just had the flu and she was feeling better. While she was taking care of us she had a relapse. My father also became sick. My mother and father both had to go into the hospital. We kids couldn't stay in the apartment, so we were taken to my grandparent's house on Boylston Street in Malden.

The last time I saw my mother was in the house in Chelsea. She had her hands out and said, 'Let me hug them.' They wouldn't let us near her. They took my mother to the hospital and she died there on October 12, 1918. She was so sick—I think she would have

OPPOSITE: SAUL,
LILLIAN AND
JULIE GOLDMAN
BEFORE THEIR
MOTHER DIED OF
THE FLU
EPIDEMIC IN
1918.

survived if she hadn't spent so much time taking care of us. She probably got double pneumonia from the flu. In those days there were no antibiotics.

I was almost five. I remember going to the cemetery with my brother. Half a million people died in the U. S. from the flu epidemic. There were 15 or 20 boxes lined up with dead people in them. My father wasn't there, he was still sick in the hospital. He didn't even know that Ma had died until they brought him back to the house. My grandparents and uncle were at the cemetery. They wanted to open up the box to have a last look at my mother and then decided not to. Someone cut a button and the lapel off Saul's coat and my coat. They pushed the buttons and lapels under the coffin lid, and buried her.

Since my father was still in the hospital he couldn't take care of Lillian and Saul and me. I had three aunts, my mother's sisters, Bessie, Sophie and Fannie. Bessie lived in New York, and Sophie and Fannie lived in Hartford, Connecticut. My grandmother sent Lillian to Auntie Fannie. Saul stayed with Aunt Sophie. I was sent to Auntie Bessie.

I was at Aunt Bessie's maybe a month or so. They had an older son and he tried to abuse me and I didn't know what to do. I went into the house and I said that he peed on me. They knew he was up to no good so they sent me right back to Boston to my grandmother's. She couldn't take care of Saul and me. She was very old at the time.

MY FATHER CAME HOME from the hospital. He couldn't take care of us and work. He sent Saul and me to an orphan home in Dorchester, Massachusetts on Cantebury Street. The Hebrew Industrial School was a home for Jewish children. My father paid the school to take care of us. Lillian stayed in Connecticut with Auntie Fannie. She was only there for six months, then brought back to live with Grandma and Grandpa.

It was the first day and we had no clothes with us—nothing. The women who ran the place took us into a bare room. It had a bed in it and bars on the windows. To make sure we were clean, they kept us there for a couple of weeks. They thought we might be contagious with a disease. My brother and I had our heads shaved. We were bald. They didn't want us to spread lice. Saul and I stayed together in that room. We weren't given any toys or books or anything to do.

We just didn't know what to make of it. We'd stand at the windows all day looking at the kids out in the yard. We didn't have contact with anyone. They brought food to us. Finally they let us out. There were about 300 kids in the home. It was a three-story building with a cellar. There were two dormitories on the second floor for big and little girls, and two on the third floor for big and little boys. Each dormitory had lines of cots. They put me with the little boys and Saul was with the big boys. They never kept us together. The dining hall was on the first floor and then they had a cellar. They had fire drills every so often and showed us how to get outside.

At the orphan home, they'd wake us at six in the morning. All the kids had to take a cold shower down in the cellar. If you didn't want to go in they threw you in. After we got out and dried off we went out to the playground and exercised. Little boys like me had one pound Indian clubs. We had to swing the Indian clubs for twenty minutes solid. Big boys had two pounders. After we got our exercise we went in and had breakfast. It was usually oatmeal or eggs. Then we'd get dressed and march to school two miles away. It took about 30 minutes of walking in line to get there.

They sent us to a public school. I was interested in learning. We did whatever was necessary, and then marched home. We marched in the morning, marched back for lunch, then marched back to school and stayed there until late afternoon. We also went into the school at the orphan home. The big boys learned trades such as carpentry, machine shop, and printing. The small boys went into Hebrew School and learned the alphabet. I started to learn Hebrew. I can remember the alphabet.

Whether it was summer or winter we marched to school. If it rained, it didn't make any difference. They'd march us there and march us back. Saul and I were going to Sarah Greenwood School. The big boys were going on streetcars to Boston Latin and other high schools.

The women who ran the orphanage were all very strict. None of us liked them. At mealtime we didn't dare make a noise. If we made some noise they'd point us out. That meant whoever made the noise had one demerit. At bedtime, all the kids lined up in front of the bunks. The women walked around us. If we had two demerits, it was bap bap with the strap—a cat-o-nine tails. We had to pick up our nightgowns for them to hit us.

There was enough food. They fed us three meals a day. I liked the hot dogs better than anything they fed

us. I used to keep one in my pocket and eat the other one. We'd march to school and I would be sucking on the hot dog like a lollipop, eating it while I marched.

There was a stone wall behind the home that circled the orphanage playground and connected to the sides of the house. On the other side of the playground wall was a 'nut' house. One day the fellows wanted apples. Apple trees grew just over the wall in the yard of the 'nut' house. The kids boosted me over the wall and then boosted a girl over. She was hanging down on the other side. I caught her when she let go. I had no bags with me and couldn't carry the apples in my hands. I pulled the apples off the tree and put them in her bloomers. When the bloomers were full I boosted her over the wall and then I got back over. The wall was about six feet high but I was like a monkey. We enjoyed those apples.

In the orphanage, we had a large hall like an auditorium. Eddie Cantor was a famous actor who came around and did shows for the kids. I sat in his lap one time. He used to buy baseball gloves and toys for us. He was known as 'Banjo Eyes.' He played in *Kid Boots*, a famous play back then. He did his own act with two or three other people from the cast. He bought bobsleds, skis and skates for us and took us to Franklin Park, near the zoo. We had meals there and shared the skates and sleds. There were two hills in the park. The bobsleds went down at pretty good speeds. We played as much as we could, but we had to go to school.

The orphanage was on Cantebury Street, which is now American Legion Highway. The orphanage changed its name to the Hecht Neighborhood House in 1922. It started in Boston, in the west end. Then the west end was torn down and it moved to the former orphan home.

Thousands of children were orphaned by the flu epidemic and the remaining family members didn't know what to do with the kids. So they were put into orphanages. Couples who lost *their* children would come to the home and look over us kids. If they liked the looks of a kid, they'd say, 'I'd like to talk to that boy.' They'd get to take us out. They didn't know where the child came from, or whether he had anybody. I hid from one lady because I didn't want her. I just wanted my mother. The woman gave me a baseball glove and a ball and she took me to the zoo. As soon as I realized what was going on, I disappeared. The next week when she came I was hiding in the boiler room.

I didn't know that my mother wasn't coming back. I used to dream about her all the time. I had one little picture that I carried in my wallet for 60 years. It's all faded. I have a picture of the three of us, Saul, myself and Lillian, when I was five, before my mother died. I always thought that I'd see my mother again someday. She was so pretty in the pictures that I had, and I thought she was somewhere where I'd be able to see her. Maybe she was in Russia, I didn't know. All I knew was that she'd disappeared. They told me she'd died but to me it didn't mean anything. I wanted to see her somewhere.

I didn't want to leave when couples would come to take me out. All those kids were like my brothers and sisters. The closest friends I made were Murray Cohen and Max Simons. My father used to visit us every week. My Uncle Iz used to come occasionally. We had other relatives that would come and visit us. They'd be unhappy that we were there, but that was it. My father took us out when I was six.

MONTHS HAD gone by and my father met a woman named Clara. She had been married before and lost her husband and child in the flu epidemic. Her family thought she and my father should get married. So they did. They had a little boy a year later. His name was Hyman. She was very nice to us at first. But once Hymie was born, she was preoccupied with him, and Saul and I were hard to handle.

My father bought a house in Malden, 12 Holyoke Street, and we all lived there. It was a two-family house. We lived upstairs. Clara couldn't handle us and we were always in trouble. There were a lot of arguments between us. If we didn't do exactly what she wanted she'd whack us. One time she locked me in a room. I climbed out the window into a tree, slid down to the ground and ran away. They found me a day later out in a swamp in Malden. I was there all day and all night, hiding. I was about six. I got a few whacks from my father for that.

I just didn't want to be there, with her. My father tried to make me call her mother and I refused. I never called her mother. I called her Auntie. She didn't like that. Saul and I had the same problem. He ran away to different places but they'd catch him right away. In Malden we went to Judson School in Judson Square, near our house on Holyoke St. I was in first or second

grade. One day I took a box of pen points from the counter at school, and brought them home. I was playing with them, and Clara said, 'Where did you get those?' I said, 'I got them in school.' She said, 'Who did you ask for them?' I said, 'I didn't ask anybody, I just took them.' She got me by the ear and towed me down to school. The teacher threw me in a closet. I was locked up for two or three hours and then I was sent home.

CLARA'S BROTHER SAM owned this big supply house on Washington Street, in Boston. Sam talked my father into starting a wig factory. We made wigs in the attic. Saul and I worked along with a girl who was five years older than us. Sam sold our father all the spindles and the hair to make the wigs. Baskets of hair came from everywhere. Pa taught us how to weave. I could weave a pretty nice wig, and I was only

THE ORPHANAGE JULIE STAYED IN WITH SAUL BECAME THE HECHT HOUSE, A GATHERING PLACE FOR JEWS OF ROXBURY AND DORCHESTER.

six or seven. We'd weave the hair and then it had to be given to a wigmaker for framing. They had no shape to them, they'd just lay on the head like a clump. As soon as we got home from school we'd go up into the attic.

We made about 20 baskets of wigs altogether. Pa said, 'I'll take them over to Sam Bernstein and sell them to him.' When he got there, Sam said, 'What are you doing?' He said, 'I'm giving you the wigs. You sold me the idea of making wigs and now you can sell them.' He said, 'No, I didn't tell you I could sell them for you.' They had a battle royal but Sam wouldn't sell them. That ended the wig business. Pa threw all the wigs away because he couldn't sell them. Sam probably could have sold them in New York in the Jewish areas. Hasidic women would have bought them. They shaved their heads and then wore wigs. I didn't want to make wigs and when my father gave it up that was great.

LILLIAN WAS STILL living in Connecticut. My Aunt Fannie wanted to adopt her and my father didn't want that. He brought her back to Malden so Clara could take care of her. That was an even tougher job for Clara. Pa couldn't get along with Clara. They were fighting like cats and dogs all the time. They would always yell at each other, about anything. If she made the eggs dry he'd yell at her. He was a tough guy and it was hard to get along with him. She would yell at him, and they had bad words. Pa was a tailor at that time. He had a tailor shop and made pants and he did all kind of things trying to make a living.

We lived in Malden for a year and they were fighting

all the time. In those days husbands could put their wives in a 'nut' house. Pa said she was nutty and he put her in the Danvers 'nut' house. I went with him on a Sunday to visit her. I didn't know she was in a mental hospital until he took me there. I didn't know where she was and I didn't care. I wasn't allowed inside. I waited outside while Pa went to see her. All the windows were barred and she saw me standing outside. The window was open. She said, 'Tell your father to take me out. You can do it, tell him to take me out.' I looked at her and said, 'What am I going to do?' I wasn't very enamored with her anyhow. No matter what I did she'd yell at me and slap me. Pa brought her home after keeping her there a while.

The second time Saul and I went to the orphanage, I was eight. We stayed there seven or eight months. Saul got very sick there. He had rheumatic fever and they put him in a hospital somewhere. I didn't know where he was. I had swollen glands, and someone at the orphanage brought me to the hospital on Humboldt Avenue in the Eggleston area of Roxbury. To get the swelling down the doctor cut my throat to let it bleed. I had adenoid tonsils and the doctor said they had to come out. He operated and Uncle Iz came to get me three days later.

I couldn't go back to the orphanage because I wasn't well. Uncle Iz was told to take me to a farm in Walpole, run by an Irishman and a Scottish woman. The home was a part of the Jewish Federation. There were six or eight kids there. They took good care of us and that's where I found my brother. I hadn't seen him for a couple

of months. We were happy to see each other again.

I made my first radio in the orphan home in 1923. We had no radio and there wasn't any television in those days. Before that, the only time we heard news was when my father told us, or when we learned it in school. I learned how to make the radio by watching somebody else. I got an oatmeal box and copper wire and I wound it around tight. I got a crystal set. I put the end of the wire on the crystal, and could pick up radio signals. I heard Jack Dempsey and Firpo fight on that radio. Firpo was a world championship fighter from Argentina. Dempsey knocked him out in the second round.

At the orphanage, I met two friends, Murray Cohen and Mack Simons. Murray Cohen and I snuck into a Sunday School to see what it was like. It was on Blue Hill Avenue. We'd skip school sometimes. We'd get out of the parade and see what the rest of the world looked like. I used to like the church and Murray would say, 'Let's hurry. They know that we're here.' I had just started to study in a Jewish class at the orphan home. Every kid, seven or eight years old started Jewish classes and then when he was 13 he would get a bar mitzvah.

PA DECIDED TO LEAVE Clara and buy a farm. The farm was in Middleboro, Massachusetts on about 50 acres of land. The land had some trees, pasture land, and fields of hay. He paid about $2800 for fifty acres, the house, the barn, and everything else that went with it. Saul, Lillian, and I went with him. Clara didn't know where we were. She thought that we were living somewhere in Dorchester. She stayed in Malden

OPPOSITE: NATHAN GOLDMAN, JULIE'S FATHER, PREPARES FOR A SWIM AT NEARBY REVERE BEACH.

at the house with Hymie. Pa had bought the farm and never told her.

A Portuguese family named Gomes owned the farm before us, and they had 22 children. There must have been 20 broken bicycles lying around the place. The well was full of broken glass and trash. They put me down there on a rope and I kept sending up trash until it was cleaned out. I was hanging in there a day at a time, cleaning it out. It smelled awful. Once we got it all pumped out and got all the mess out, we had good fresh water. We had a water pump outside. Eventually we ran a pipe into the house and had a water pump inside. We put a bathtub in the spare room. We'd fill it with water which was first heated on the oven in a copper washtub. When we pulled the plug in the tub the water went into a pipe and then into the ground outside the edge of the house.

Lillian and I went to school in Middleboro. Lillian was eight and was going into the third grade when we first moved to the farm. My brother didn't go to school anymore, he stayed right there on the farm and worked. I was twelve and Saul was thirteen. My father worked in the clothing business in Boston so my brother was left all alone on the farm to take care of it. He was only a kid himself.

Clara had a woman's modiste shop, selling corsets and bras and things on Winter Street in Boston. Three months passed before Clara found out about the farm and got in touch with my father. She wanted to live with us. Her plan was to hire a nanny to take care of us. Then she and my father could go to work in Boston together. Pa agreed and Clara hired a nanny, an African woman. The nanny was very good. She cooked and made things that we had never had. We had a big blueberry patch, and we'd pick a box of blueberries in five minutes, the size of a half-dollar. The nanny made us blueberry pie. She took care of the cooking and that sort of thing, but she didn't do anything on the farm. She took care of the house while my father and Clara went to Boston and worked. It wasn't long before Clara and my father had some disagreements and fought like cats and dogs again.

IT WAS 1924 when Lillian was in third grade. On our way to school, Lillian and I would bring Hymie to kindergarten. We rode the street car that ran a mile and a half from us, from Taunton to Middleboro. Clara was home one morning after Pa had left for Boston. She said, 'I'm taking Hymie to school today so you won't have to bother with him.' I was happy about that.

Lillian and I took off, got on the street car and went to school. We came home just before dark and didn't see anybody around. There were no lights on in the house. I opened the door, and there was no furniture. I walked into the next room, no furniture, and the next room, no furniture. The house was stripped bare. There wasn't anything left in the house, not even a match. The only thing I found was my gun. It was hidden in the eaves with my bullets. She didn't know where it was.

There were no dishes. There was nothing left. Nothing. Not a kerosene lamp, not even matches. Lillian sat down in the middle of the floor and she started to cry. She was just a young girl. I didn't know what to do. I took her to the farm next door, owned by a Polish family named Smarsh. I brought her over there and told them that I'd wait for my father. I sat on the stairs at the stoop of the farm. I sat for over an hour and then I heard the old car chugging up the road.

It was after dark. I knew it was his car because I could tell by the whine of the engine. It was an old 1920 Chevrolet. I was sitting on the stairs and the lights picked me up. He swung around to the small garage and drove in. He walked up to me and said, 'Why are you sitting out here? Why aren't you in the house?' I

said, 'Well you better come in the house and look.' He said, 'Why aren't there any lights?' I said, 'There are no lights. Go in and look.' He found some matches in the barn. He lit up a match and he looked around the house and he went crazy. He grabbed my gun and the bullets. He said, 'I know where she is. I'm going to go and kill her.' And he jumped in his car and took off.

I ran to the Smarshes who had a telephone. We called the state police in Bridgewater and I told them the story, I told them my father was going to Boston to kill Clara. They said don't worry, they'd stop him. And they did. They caught up to him on the main road from Bridgewater to Brockton. They took the gun from him and told him to go home and cool off. He eventually went to court and got a divorce. The judge ordered Clara to give all the furniture back. We got a few things, but not my mother's piano. There was a total separation of the two of them from that day.

WHEN MY MOTHER was alive, my father studied piano at the Conservatory of Music, in Boston for a few months. He was very good as I understand it. When my mother died he stopped playing. We were in the house in Middleboro, and we asked him to play it for us. There were two or three songs that he remembered like *Mother's Prayer*, very touching songs, back to back and then he'd sit there. When my mother died, I think that made him a bitter man. Clara stole that piano and when he said he didn't want to see her again she gave the piano to her nephew, Leonard. It was Leonard Bernstein. He said that the first time he played a piano was the one that his Aunt Clara gave him—my mother's piano.

I didn't get along with Leonard. I kicked him in the butt once and knocked him down the stairs. I was seven or eight years old. He was a scrawny little kid. I think I was with him on Revere Beach a couple of times. I never saw him after that. He was in pictures, but I never saw him.

WE HAD TWO COWS and a horse. Saul had to milk the cows. A good cow gave fourteen quarts of milk at a time. Inside of two or three days we had a lot of milk collected in a milk can. A milkman would come along and pick it up. The milk sold for 10 cents a quart.

FRONT ROW: LEONARD BERNSTEIN, JULIE, SAUL AND LILLIAN. BACK ROW: LEONARD'S MOM (JENNIE), CLARA, NATHAN AND HIS MOTHER, GITTEL.

We had chickens so there were plenty of eggs. We sold them to an egg man. We had a hay field for the cows and a garden where we grew vegetables. The potatoes were stored in the cellar where they would keep all winter. We could use them as needed. In the spring the potatoes sprouted little eyes. We cut the potatoes up with two eyes per piece and planted them in the ground.

At that time my father had a truck and he was a peddler. He bought clothes in Boston and sold them around the farms and in the city. He'd go door-to-door with his truck and people could buy new clothes from him. He sold pants and shirts. My brother and I stole eggs from our farm and sold them to a customer of my father's. One day the woman said to Pa, 'Those are wonderful eggs that your son brings over. Could you get me some more?' We were selling eggs for 12 cents a dozen, a cent a piece. But Pa caught us. We caught hell for that.

I made my own ice cream and cottage cheese. I'd put cheese in a pointed bag on a board. I'd put another board over it with rocks on top of it and squeeze the water out. When the water was all out, it was cottage cheese. I made ice cream once in a while. We had a little churn. I'd put rock salt and cream in the bucket outside and start churning it with ice. Eventually it got churned up and we'd have ice cream. There was always some buttermilk left in there, that was nice.

Saul and I never took a bath except when we were forced. Aunt Rose and Uncle Iz would come over. They'd throw us in the bath and wash us. They had

three kids then, Helen, Sonny, and Sylvia. Aunt Rose took one look at our hair and said, 'You've got nits in there.' She shaved our heads and washed us with lye to remove the lice.

We had no electric and we had no heat other than wood. We had kerosene lamps, and lanterns that we'd carry into the barn. We had to be careful not to burn the place down.

The refrigeration was in the cellar, but we had an icebox in the kitchen. A cake of ice would last a day and a half or so, then we'd have to buy another block. In Middleboro, Never Touch Pond was right near the elementary school. In the wintertime men went out and cut long blocks of ice. They had a tram that would bring the ice into the icehouse on the side of the pond. They put sawdust down and made rows of ice. They used saws to cut the ice by hand. The ice man would go there and fill up his wagon. He'd come around to the

NATHAN GOLDMAN ON THE FARM WAGON WITH SAUL AND JULIE.

farms and sell us as much ice as we needed. If it was a warm winter there wouldn't be much ice. Then they'd have to import the ice from somewhere else.

For meat, we had chickens, but my father used to buy hamburger for Lillian. She had the hamburger or the steak because she was a girl. He told us boys didn't need it, girls needed it. Sometimes he'd leave the icebox unlocked and we'd steal some food out of there. We always had apples and all the local stuff. My brother built a little bog in the low area of the farm, and grew cranberries. We didn't sell them, we just ate them ourselves.

We had eggs, vegetables, flour and cabbage in the cellar. It wasn't a big cellar but we had a lot of stuff in it. We had a lot of apples so we bought a cider machine. We'd grind the apples up and make cider. But we never had soda to add to it, to keep it from turning sour and becoming liquor, or sour mash.

LON MELIX WAS AN INDIAN who lived behind the farm. He lived in a shed with his horse and he took care of the cranberry bogs owned by Ocean Spray. He'd come over to our house at night. My father would take out two pitchers of cider that had turned in one of our 50-gallon barrels. They'd drink it all night and play checkers. Not for money. Neither one of them had money. The Indian would tell a story, and we'd hang on and listen to the stories. They were pretty sharp. They were friends always. The only thing Lon Melix owned was a horse— the shed belonged to Ocean Spray. His horse was on one side of the shed and he slept on the other side.

MR. BENSON MANAGED Ocean Spray and was the commissioner of Boy Scouts for all of the Cape area, Taunton and Middleboro and other towns. He picked up kids from different farms in his convertible and we'd have a meeting. We were called Lone Scouts. Each kid was a scout trooper by himself on his farm. Benson would pick up as many kids as he could fit on that car. There'd be seven to ten kids hanging all over it. It was a Packard Roadster, a fancy car. He took us to an Ocean Spray building and taught us to make knots among other things.

I was in the scouts for about three months. One time Benson picked us up and it started to snow. He got stuck and blew a tire. His car had thin tires. All the kids got together and we changed the tire. It took a couple of hours, and then he drove us home. Pa got home before I did that day. When I got in he said, 'What were you doing?' Pa never knew I was a Lone Scout. I'd get my chores done but he never knew I wasn't at the farm. I came home and he blew his cork. Up in the attic I had a little room. On the wall I had a skunk skin, a fox skin, and a chart I made as a lone scout—seven or eight pieces of paper in all. He went up there, opened up the window and threw everything out into the snow storm. I never saw it again, it just blew away. That ended my lone scouting.

SAUL AND I HAD A SKUNK trapped. We shot him but he crawled into a hole. Our dog, Prince, was digging at it and the skunk sprayed him. We got the skunk out and peeled off his hide. Every time it

rained the dog smelled like a skunk. We couldn't get the smell off him. When he dried out he didn't smell too bad. He was a wire-haired terrier, a tough dog. He would shake his head all the time. We didn't tell Pa about the skunk. He would throw the dog out the door every time he got wet. Then Prince would dry off and he wouldn't smell.

Pa loved the dog. He picked him up in a dog pound in Boston. We used to have a chain that went from the shed to the barn. He'd run back and forth and that dog knew every foot of land that we owned. Anyone who'd put their foot inside that land would lose it to Prince. He'd be silent and then the minute a stranger got his foot across, he'd growl.

The neighbor had a dog for 15 years and the dog was sick and crawling. It was not as big as a St. Bernard but a large dog just the same. The neighbor came over and said, 'I have to shoot him and I can't do it.' I said, 'Well I'm not going to do it.' He said, 'I can't do it.' I took my rifle and I walked the dog into the woods and I shot him. The dog knew I was going to shoot him because I took my gun. I had to bury him too. My neighbor wouldn't even bury him. I felt bad.

Saul and I used to shoot birds all the time. He was a hell of a shot. He could kill a bird on the fly and it would come down on the top of your head. We didn't eat them, we'd just shoot them. We were always shooting something. We used to take a bottle and a mirror and shoot backwards, shooting the bottle off the fence.

I shot the mailbox full of holes and Hymie watched me do it. The next day I went out to get the mail. It had rained and the mail was all wet. Pa asked 'Who did this?' I said, 'Oh I don't know, some kid going by.' Hymie took the gun and he shot out the window toward the barn and Pa caught him. He said, 'Well if Julie can shoot the mailbox full of holes, I can shoot the windows out.'

Pa couldn't give me anything. I probably saved up and bought my gun from some farm kid. I didn't go to a store for it. I could buy things from other farm kids, on occasion. Every farm kid had a gun. Some of them had shotguns. There was a farm next to us owned by the Races. They had three boys, and Saul and one of the boys were pretty friendly. The boys came over and had some apples with Saul one day. I told them, 'I don't want you in here.' I didn't like them. Saul and I had a hell of a fight. He took a hammer and hit me on the head with it. I caught it before he buried it too deep.

My cousin Sonny used to visit. Sonny lived in Malden as a kid. He and my brother and I would get together and cause trouble. He'd stay for a week or two on summer break. The three of us would go to the store owned by a man named Gary. He only had one long tooth in the front. We used to call him Fang. One of the kids kept him occupied and we stole cigarettes and cigars from him. We got home and started smoking. I had to shovel a lot of the manure out from the barn and put it on top of the wagon and take it out in the field. I felt sick from the smoke and the smell of the manure. I went into the house. As I heaved out the window Pa came in. He said, 'What's the matter?' I said, 'I guess I'm a little sick.' He found out what we were doing and said, 'You don't steal those cigarettes or cigars anymore.' I wasn't going to smoke again after that because I got so sick, and I never did.

Gary was a good old fellow but we really gave him trouble. We were just fooling around being kids. On Halloween, Saul and Sonny and I took Gary's wagon apart. We climbed up on his roof with the parts of his wagon, piece by piece, and put it back together again. We put a bucket on his smoking chimney. We got down off the roof and were watching the front door. The smoke backed up in his store and he came out raving mad with tears in his eyes. He pulled a shotgun out and he shot at us. It's a wonder we didn't get killed. We had to take the wagon apart and get it down. He didn't tell Pa, but he wasn't too happy with us.

BOOKS WERE MY OUTLET from working my tail off on the farm. I would read any book that had printing. I read Nick Carter and Edgar Rice Burroughs. My father hated the books I picked. He said, 'You won't learn anything from that.' I used to go to the library. My brother and I were allowed seven books each for two weeks. I'd get 14 books and I'd struggle home with them. I'd hide books like *Tarzan*, and *The Riders of the Blue Sage*, cowboy stuff. It didn't make any difference. It could be a love story. I wouldn't understand it but I'd read it.

When I was in the seventh grade, one of my teachers, Mrs. Kidd, befriended me. She asked me about my vocabulary. She said, 'How do you have the diction you have? You don't have a mother, who teaches you?' I said, 'I read by myself.' She said, 'I want you to read constantly. Reading is the best thing you can do. Read, read and read.' I kept getting the library books to read. Pa said I should be reading law books. I was reading Zane Gray and Tom Swift and Frank Merriwell, detective stories and Civil War stories. I'd read them all but he would never see me. I would read in the woods, or in the loft of the barn. I'd light a candle under the blanket in the attic and I'd read my books. My father never wanted us to read. That took time away from our work on the farm.

Pa came home early one day and I wasn't around. I was in the pine grove reading. I looked around and he was standing there, looking down at me. He said, 'I knew you weren't doing your chores. You're here reading.' And he gave me hell. He took the book away. It was given to me for a book report in school. It was fiction. He said, 'You don't get those kind of books in school,' and he took it and threw it in the fire. Later on, he had to pay for the book. I hid my books most of the time in the loft. How he ever found out that they were there I don't know. He put a pitchfork into the hay and out came a book stuck in it. He gave me hell for reading those books. I never finished my work because I was always reading or playing sports.

One time I was playing football and didn't get home in time to milk the cows and give them water. A cow was mooing, and Pa asked, 'Did you give her water?' I said, 'Yeah.' He said, 'What's the cow mooing for,' and I said, 'I guess she's happy.' He got the cow a bucket of water and she sucked it all in. Then he gave her another one and she drank that too. He said, 'That's why we're not getting any milk.' We'd get 14 quarts of milk in one milking. He said, 'You never gave them any water,' which I hadn't done. He hit me with that bucket and when I came to the stars were out. I had been unconscious for quite a while.

Pa never knew that I played sports, ever, or that I was captain of three teams. I was captain of the football team, the track team and the baseball team in my school. When I started to play football and baseball I learned how to fight. I'd be playing along and I'd whack somebody. I got to be a pretty good scrapper. I was a hothead. Pa never showed Saul or me any affection. I don't remember him ever touching us except to hit us. I don't remember any happy times when I was a kid. I was just happy to see cousins and that. So was Saul. I'd talk to my father about Sonny. Sonny's father would take

him to baseball and football. He ran in many races and we'd follow him. I said, 'Why can Sonny do that,' and my father said, 'You're different, you've got no mother.'

One time I had to reshingle a barn. I used 14,000 shingles on that barn. They came in packages of 250 shingles per block. I had the football team come and help me put on shingles in the dark. They came after practice to the house so that I could get my chores done. I struck the line with the ground not realizing that the barn was sitting on a slope. When Pa came out in the morning and saw all the shingles facing downward, he said, 'The barn is sinking.' When he got close he realized I had put the shingles on cockeyed. He raised hell. He tore the shingles off with a hook and he gave me a few whacks for it. He said it was no good. I asked why. He said, 'It's no good because you didn't have your heart in it.' Well I didn't. I only did it because I was made to do it.

All three of us cooked meals and cleaned up around the house. We made eggs and hamburger. It was quiet during meals. My father hit Saul with an egg one time. He said, 'I told you I wanted a soft-boiled egg and you made it hard.' He threw the egg at him. It hit him in the head and the egg was soft. It bounced off his head and went into my hat hanging on the chair. I had a hat full of egg. We did all the laundry too. We had a washboard that we stuck in the bathtub to wash our clothes by hand. We were pretty busy kids, not that we liked it, but we had to be that way.

Saul would hang around in the mornings with me at school, then bum his way back to the farm and do his chores. When I got home we had to wash the dishes and clean the house and take care of the cows and the horse. It was about a mile and a half walk to get to the trolley line, then the trolley went into Middleboro near the school. I never had a bike. Anywhere I went I walked. At night, my father would just sleep. He seemed bitter as heck. He was always criticizing us about something.

WE WERE REMINDED all the time that we were Jewish. Irish kids called me a Jew. There was a Jewish family named Green about eight miles away. They owned a big dairy farm. They had a lot of things that we never had. There were also two Jewish families in the city, the Freedmans and the Bermans, both junk dealers. The Bermans had a son, Ike. They used to buy junked cars for $20 a piece, and they'd sit in the yard and try to sell them. If they couldn't sell the cars they'd disassemble them and sell them for junk. I got to drive locomobiles and Packards all around the junkyard.

Pa might of lost his religion when my mother died. My grandparents were very religious. I was just a little kid and all of it rolled off my back. Grandpa used to come in to town and visit. He'd ask me to do things and he acted mad all the time.

Auntie Fannie came up to see us at the farm from Connecticut. Once every six weeks she'd visit and give us some money. Five bucks was a lot of money in those days. She bought me a new pair of dungarees—jeans. They used to go for two bucks. I had an old pair that was torn in the crotch. I didn't feel like wearing them, so I put on the new dungarees. Grandpa said, 'You can't wear that.' I asked why, and he said, 'That's for Saturday, that's the day you have to dress up.' I said, 'No way.' He chased me down the cellar. He was running around yelling at me. 'You bum,' he said, 'you'll be nothing but a *balagula*,' which was Yiddish for truck driver. His head hit one of the beams in the cellar, and it rang like a bell. I laughed so hard I fell on the floor. He couldn't get a hold of me. I think he hated me.

Saturday was the Jewish day of rest. We never observed them. My father never even told Grandpa that we had ham and bacon. If he came over and found out what we were eating we'd be in trouble. Saul and I were never given a bar mitzvah. I wasn't interested in religion. There was nothing in the house that made me know that I was Jewish.

We never celebrated Hanukkah. We never had any traditional Jewish teaching until kindergarten at the orphanage. We eventually learned Yiddish from listening to our father and grandparents talk. They were saying things they thought we couldn't understand but Saul and I could understand it.

Mrs. Kidd asked me one day if I'd want to be adopted. She said, 'You have no mother and I'll take good care of you.' Her husband was dead, and she had two girls. She thought that I was pretty bright. I said no. I said, 'Don't tell my father, because I'll get in a lot of trouble.' I didn't want her even though she was a nice lady. I was 13 years old. I didn't want to be adopted in the orphan home, and I didn't want to be adopted by her.

I WAS IN SIXTH GRADE and Saul was still working the farm. Saul used to come and visit me at the school during recess, in the morning and in the afternoon. He'd be sitting there waiting for me. The teachers saw this and they started to ask questions.

The Never Touch Elementary School was next to the ice pond. Mrs. Jones was in the elementary school. She found out Saul had never gone to school. She told Saul, 'If you go to school we'll let you hang around here.' He said, 'I won't do it.' She made him go to school and he was sharp. He made everything up and got through the sixth grade in six months. We never told my father until he got the diploma. Saul showed it to him and boy, he blew his cork. My father said, 'Now I know why the work hasn't been done. You haven't been here.' Pa said he couldn't go to school or into town any more. He had to stay at the farm and work.

A teacher from the school told my father that all the kids in Massachusetts had to go to school. The school system wanted to take my father to court but it was too late. Saul was already gone. Saul was 15 when he moved to Boston and worked for a time-clock company as a installer of card machines. He lived in a room somewhere in Boston and never went beyond the sixth grade.

One day after Saul had left, Pa didn't take his car to the train. It was a 1920 Chevrolet coupe and it had a funny whine. When I'd hear that whine I'd stop reading. I'd put the book down and start chopping wood. But this day, I wanted to learn how to drive. I was 13 or 14. I backed the car out of the garage, onto the dirt road in front of the farm, and headed toward Taunton. I got

two or three miles down and I couldn't turn around. I backed the car all the way up to the farm, then came in and headed towards the garage again.

As I got into the garage, I missed the brake, stepped on the gas and knocked the back wall down. I knew I was going to catch hell after that so I tried to fix it. I got the wall hooked onto the front of the car, got the top of the wall with a rope, and I pulled it back up. I nailed it back into position but there was a space in the corners. My father went to get the car the next day. He said, 'The barn, there's something wrong with it.' But he didn't know what. Lillian knew I had taken the car. She was blackmailing me. I had to wash the dishes for her. She said, 'I'll tell Pa,' so I did everything. I wouldn't give in anymore and she told Pa. He belted me a few times.

I WAS 12 WHEN my father decided he couldn't handle me anymore and he took me in his car and left me off in Boston. He gave me a dollar and said, 'You're on your own.' I wondered what I would do. I remembered that Clara had a modiste shop on Winter Street. I went there and told her I'd like to stay there and sell papers in the corner stores. She said I could stay there but only temporarily. I only lasted one night. The women were coming in there and changing their clothes. I was ogling them because I was a young boy. Clara said, 'You can't stay here.'

I was able to handle a streetcar pretty well. I used to go from the orphan home to Malden when I was little. For a nickel I could go from Dorchester to Malden. First I'd take the streetcar and then I'd go on the underground. Then I'd take the elevator to another streetcar to Malden, all for a nickel. I knew how to get around. I got into the underground on Winter Street and I went to my Uncle Iz. He lived on Warren Avenue in Malden. He had three children at the time and my Aunt Rose was pregnant. He said, 'I have no room for you here, but I'll let you sleep in the hall and use the second floor.'

I told Iz what had happened with my father and he was very unhappy about it. He got me a job in a bowling alley setting up pins. I got three cents a string. In those days they used to charge ten cents to the customer. Everything was done by hand. My fingers were bent and gnarled within a month. You'd have to pick up four pins fast, setting them up two in each hand. I sat down between two alleys and I had to watch out. When the balls would come flying, they were small bowling balls and they'd fly like bullets. I had to set up 33 strings to make a dollar.

I slept on a cot in their hallway and I didn't have to pay any rent. If I made six dollars in a week, I'd give three or four dollars to Uncle Iz. Finally Uncle Iz got good and mad. He went to my father and had a hell of a fight with him. My father said, 'I wanted him to see how bad it was out in the world so that he would come back and do what I told him.' He was trying to teach me a lesson. I went back to the farm. I made up the three months or so that I lost in school, and I got some fair marks.

I became friends with Ike Berman in eighth grade. When I graduated junior high, I had no suit to get on the stage. I asked Pa, since he was selling clothes. I asked if I could borrow a suit just to wear for the night. He said no. He said I wasn't entitled to it because I hadn't done the right things. He said, 'You're a bum and I'm not going to give you anything.' Ike was a tall kid, and he had an extra suit. I looked like Digger O'Dell, with a black suit hanging off my body. I got on that stage and I graduated wearing a suit. Pa didn't go to my graduation.

I WAS LOOKING FORWARD to high school. I went and told Pa. He said, 'You can't go to high school, you have to stay here and take care of the farm.' Pa felt it was all right for me to go to school until Saul left. Then it was up to me to take care of the farm.

I gave it some thought, never told anybody anything, and that day I bummed my way to Boston. I went to the Marines recruiting office. In order to join I had to have a medical examination. The doctor looked at me and said, 'Hey, grow up.' I said, 'But I'm 18.' At that time I was 14. Then he said, 'Well, we can't take you anyway, you got a heart murmur.' I didn't have one, he just wanted to get rid of me. He had one look at me naked and knew I wasn't 18. I bummed my way back to Middleboro. My father never knew I was gone. The next day I bummed my way to Providence, Rhode Island and tried another recruiting office.

Back then, the speed limit was 35 miles per hour to conserve tires. It took me most of the day to get there and get back. I wanted to go into the Cavalry because we had a horse on the farm. I could ride pretty well. They had no room in the Cavalry but I could join the Field Artillery where they had horses. I said, 'Well, yeah that sounds good.' The man said, 'How old are you?' I said I was 18. I was given a medical and told to get my father's permission. The consent age was 21. I said, 'My father isn't home.' The man asked, 'You got a telephone?' I said no. He said, 'Take this paper and have him sign it.' The recruiting office sent a telegram to my father but Western Union couldn't deliver it to the farm. I headed back to Middleboro and went to the Smarshes, next door.

I called Western Union and asked if they had a telegram for Mr. Goldman. They said yes. It read, 'Your son Julius Goldman wishes to join the Army. Is he 18 and do you give your consent.' I said, 'This is Mr. Goldman speaking and my son is 18 and I give my consent. Would you please send the telegram back.' The next day I bummed my way back to Providence and said, 'Hey, didn't you get a telegram from my father? I don't have to get this paper signed.' The man said, 'Yes, we got the telegram yesterday afternoon. You're in.'

They swore me in. I was on a train that afternoon headed for Ft. Ethan Allen in Burlington, Vermont. Being in the Army was a revelation of what the world was like. I was looking at everything with big eyes. When I arrived in Vermont, they shaved off all my hair. I had a mop of hair. They had horse clippers and a guy cranked it and ran it over my head just like he was clipping a horse.

Next I was sent to the supply sergeant. I had a $150 clothing allowance to last me three years. Fifty dollars a year was more than enough if I took care of my clothes. I got so many pairs of shoes, socks, pants,

different kinds. I got three duffel bags full of clothes and shoes. I had more clothes than I ever had in my whole life. They showed me my bunk upstairs, with a foot locker and a wall locker. I put all my bags on the foot locker and went to the john.

I went down to the cellar. It was a three story brick barracks. I went into the toilet. A sign above it read 'Gonorrhea only.' I didn't know what it meant. A guy came in and said, 'How long have you been here?' I told him I had just gotten there. He said, 'You got the clap?' I said, 'No!' He said, 'Get out of there.'

When I went back upstairs I couldn't find my clothes. I saw the bunk, I was sure it was my bunk. I saw the foot locker but there was nothing there. I wandered around.

I found a corporal named Najolski, a big Polish guy about six feet tall. I was 5'8" and I weighed 129 pounds. I said, 'I can't find my clothes, where do you think they are?' He said, 'Did you mark your name on them?' I said, 'No, I got all my stuff about 30 minutes ago.' He looked down his nose at me and said, 'That'll teach you to put your name on everything before you put it down.' I said, 'okay, give me my clothes back.' He said, 'You lost them.'

I didn't get a thing back, not even a handkerchief. I had to go back to the supply sergeant and tell him I'd lost everything. He gave me hell. I was issued another full set of clothing and I came back upstairs and marked my name on every single thing that I had.

Five o'clock came and everyone got ready for Retreat. Soldiers had to get dressed in uniform and polish their shoes. Then we were to stand on the parade ground at attention while the flag was lowered. Everybody was out of the building before me. When I came out, I was wearing a little beaked hat like General Pershing. The leggings I had for field artillery were leather on the inside and canvas on the outside. The easiest way to put them on was with the canvas on the inside and the leather on the outside.

The top sergeant was standing in front of the parade. He looked at me and said, 'Who the hell are you?' The whole parade ground was standing at attention listening. I said, 'I just signed up, I'm in this outfit. Battery B, 7th Field Artillery.' He said, 'Get the hell out of here you freak.' I didn't know what to do, I almost started to cry. I was the only little kid there and my clothes were on backwards. I had tags all over me, just like in a comedy. I ran inside and the guys all started to laugh. I soon found out that I had been issued World War One clothes. I was supposed to have them altered by a tailor and get a campaign hat. I had to put the leggings on with the leather on the inside.

They were just making fun of me. I learned the hard way. I was getting $21 a month so anything that had to be altered I had to pay out of my pocket. If you bought any other special clothes they had, like gabardine peg pants instead of the woolen World War One pants, you had to pay for it. I bought stuff that would bring me up to date like all the rest of the guys.

They assigned me to be a cannoneer. I'd ride on a cannon while they did maneuvers and fired shells. After two or three months they gave me a pair of horses and put me in a swing. It took six horses to pull a cannon. The middle horses were swing horses and the heavy horses close to the cannon were the wheel horses. I rode the horse on the left, the near horse. The horse on the right was the off horse. Occasionally you'd switch them if your near horse didn't get enough exercise. The horse they gave me had a mustache. I used to wax it for him and it would stand out straight just like Czar Bill.

I kept getting Stable Police or KP. I was always doing something wrong. The drill sergeants were always yelling at me if I wasn't in synch with the others or if I laughed at someone. When we were learning to march and drill I had laughed because the fellow in front of me was marching like he had one leg in a hole. I got KP one time for a week.

I was peeling potatoes and washing dishes at five o'clock in the morning. I'd peel barrels of potatoes and onions, and wash dishes. I'd do it all day, morning, noon and night, and finish after everyone else. If I broke a dish I had to pay for a dozen so I didn't break many dishes. When I got my second KP I swapped with a guy who had stable police. I said I'd rather shovel manure because I liked the horses. We had 100 horses in the battery and I took care of one stable of horses with a couple of other guys.

My battery commander, Captain Loaf, called me in one day. He was a little fellow who got his commission in World War One. He said, 'Goldman, I hear that you're not 18.' We just got a notice that

someone from your school said that you were only 14 or 15 when you came in here.' I said, 'No, I was 18. I'm 19 now.' He said, 'Look, you aren't going to fool me but I'll table this. Do your chores here like a regular soldier every morning. I'll make sure you go to school in the afternoon in town. I want you to go through high school.' I said, 'I've already gone to high school.' He said, 'Look, you have no diploma yet, you're not fooling me. When you finish high school, if you get good enough marks we'll have you appointed to West Point.' I said, 'I'll buy that.'

Captain Loaf called me in one day and said, 'You're Jewish?' I said yes. He said, 'Don't you want the holidays off?' I said, 'What holidays?' He said, 'Yom Kippur and Rosh Hashanah.' I looked at him and I didn't really know what he meant. I said yes and he told me I could go in town to the synagogue. I went in town to the theater.

About a month later was Christmas. He said, 'Everybody worked on your holidays. Now you have to work on Christmas and New Years. You're Jewish and the other guys are Catholic and Protestant.' I got KP and Table Waiter and every other damned thing for that Christmas season.

I did all my chores in the morning—that was a lot of work, shoveling manure, taking care of my saddle and my horses. I was going to school, and everything was going nicely until Captain Loaf called me. 'We have trouble,' he said. 'The woman from your school complained to Senator Walsh.' Walsh was a famous old senator of Massachusetts. 'Walsh called the War

Department to request that they find out where you are and what you're doing. This lady knows that you're not of age.' I knew it was Mrs. Kidd, the teacher who had befriended me.

I had written her a letter when I got into the Field Artillery and told her where I was. She thought that I shouldn't be in the Army. She kept writing the military and telling them that they must send me home. The Captain said there was nothing he could do now, except give me an honorable discharge. He marked my character as excellent. I was given an honorable discharge November 1, 1928. I had a gabardine uniform, a campaign hat and about $16 in my pocket. I took the train back home and I went to the farm.

When I left the farm to run away to the army, I saw Prince on his big boulder in the center of the yard. I used to sit on it and occasionally pat him. He was sitting on that rock when he saw me leave. I said, 'Stay there' and he stayed. I had been gone for a year or more. They said Prince sat on that rock until he got sick—he almost died. When I came back he didn't want to know me. Pa got a German shepherd, who was bigger than Prince. One day they both got loose and ran away. The dogs went to the next farm and killed all the chickens. I'm sure it wasn't Prince that did it, it was the other dog. A car came around and the neighbor knew about Prince but he didn't know we had a Shepherd. Pa never admitted that his dog killed the guy's chickens. He got rid of the shephard after that.

When I came back my father had electricity. He

OPPOSITE: FORT ETHAN ALLEN IN BURLINGTON, VERMONT, 1928. JULIE WORKED AS A CANNONEER IN THE 7TH FIELD ARTILLERY CAMP.

had bought a battery-powered system with a little gasoline motor and a generator. Acid-cell batteries were lined up on a bench in the cellar. When we started the attached motor, it would charge the batteries up and we had light. The lights flickered occasionally from this homemade system. Eventually they brought power lines up the road and the farm had electricity.

I saw Mrs. Kidd and Mrs. Jones again. They had been very nice to me. Mrs. Jones was also a teacher who wanted to take care of me. There wasn't anything they could do for me then. I don't know how long I lasted at the farm. I just couldn't take it and I headed for Boston. Lillian stayed with my father and Hymie.

I went to my grandfather on Boylston Street in Malden and he wouldn't take me in. I found a room on Hazelwood Street in the same town for $3 a week. My Uncle Iz got me a job on a soda truck. I was lugging soda for Y. D. Empire Bottling. I was a strong kid and could do the work of two men. I was lugging four wooden cases of soda with 24 bottles to a case, sometimes up two or three flights of stairs. They thought I was so good they let me become a driver. I was 17 years old. I never had a driving lesson in my life. I went out and took a driving test and I passed it.

MY FATHER LEFT the farm and moved to Woolson Street in Dorchester, Massachusetts. It was mostly a Jewish community then. Hymie went back to his mother, Clara. Lillian graduated from Middleboro Memorial High School and went to Newark, New Jersey to go into nurse's training. While in Newark she was called to the site of the Hindenberg accident. She helped determine the sex of burn victims. She ended up working in a sanitarium for tuberculosis. Two years later she moved back to Woolson Street where she met and married Sid Pliskin in 1942. They were fairly young at the time.

Clara was remarried but Pa wasn't. My father met a woman named Zelda Stein from New Jersey. They got married and he wanted to come home to Boston to work. Her family wanted him to stay there and open a tailor shop. He said no, he was going home, and went. They had him arrested and threw him in the can. They said he was deserting her. He got out of jail eventually and came home. Zelda followed him up here. That marriage lasted until Pa died in 1966 — nearly 27 years. In 1939 they had a daughter, Dolly.

I MET SOME FELLOWS in Malden one night. Irving Parkin said, 'I'm going to see a girl on Hazelwood Street, want to come over?' We went into a house on my street. There were five girls and they were playing a little phonograph. They were dancing among themselves when we came in. I met Ruthie Zalko and her friend Sally Lerman. Ruthie was Irving's girlfriend and he eventually married her. The girls tried to teach me how to dance. They wanted me to dance with Sally. I looked around and I said, 'No, I'll dance with that one.' Her name was Florence. She was Sally's sister and she showed me how to dance. I had two left feet. After about two hours of dancing to the squawking phonograph we went outside. There was another girl out there. She wanted me to take her home. I said, 'No, I am taking this girl home.' Florence lived across the street, one house down. I came back a couple times that week to see her. Florence was 16, and I was 17. From then on we were inseparable.

We both lived on Hazelwood Street. Florence lived at #20 and I lived at #5 in a rented room of a three-decker. I got along with everybody in her family. They were very gentle people. But at first, Florence's parents thought I was Irish. I was invited to their house for dinner. She brought me in the house and her aunt said in Yiddish, 'He's not Jewish, throw him out.' Florence immediately said, 'No, his father lived in Malden and his grandfather still lives on Boylston Street here.' They wouldn't believe her. They said, 'His name is Goldman, he's not an Irishman.' I looked like an Irishman. I didn't look like a Jew. I didn't act like one either. I'd swear and fight at the drop of a hat, I'd clobber anybody.

My grandfather used to go around to houses and sell eggs. That was the only thing he ever did. He prayed the rest of the time. He was a respected person in Malden. The Lermans found out who he was. They checked it out and then they were okay with me.

Florence was the first one who ever was considerate to me, that ever thought of me. She wanted to keep me from getting into trouble, which I was doing all the time. I bought my first car, a Nash Roadster. It was a dark green, old clunker. At that time I was driving a truck and making $18 a week which was pretty good. It was 1929 during the Depression after the big stock market crash. Things were pretty tight. No one had any money. Food was very cheap. A steak sandwich was about 20 cents, a ham and egg sandwich was 15 cents. Spaghetti was 15 cents. If you had it with meatballs it was a quarter.

Everything was very, very reasonable compared to what it is today but if you weren't making any money, a quarter was a lot of money. I just lived in one room and bought all my food out. I used to eat occasionally at Florence's mother's house. Even though they were poor there was always room in the pot for another person to eat. I was the other person. Her mother used to make wonderful food. Her father was a baker so there was always a lot of bread. Florence had five brothers and two sisters. I met all of them.

I was always broke because I was giving people money. If anybody needed a buck I'd give it to them. I only had a pair of white flannels, a white shirt and a white sweater. Other than that I had dungarees. That's all I had in the world, yet I was making enough money to have clothes. Florence tried to stop me from giving all my money away and spending it on everybody. Florence and Sally had friends in Lewiston, Maine. On a Labor Day weekend we drove up to see the Bakers. The car broke down on the way home. I didn't get back to work and I got fired. Labor Day was the busiest day of the year. I was supposed to be delivering two truckloads of soda.

I HAD NO PLACE to work, and nobody would hire me. It was the height of the Depression and the car had a lot of problems. Florence got me a job with her father in the bakery, lugging flour and soft coal for the bakers. We'd lug the flour in and the bakers would stoke the oven. Working in the bakery was terrible. I only got $7 or $8 a week. I saw a Jewish slaughterhouse, Goldstein and Borokauf, across the

FLORENCE LERMAN WAS 16 WHEN SHE MET JULIE AT A FRIEND'S HOUSE. FROM THEN ON, THEY WERE INSEPARABLE.

street. They killed chickens and sold them to the Jewish trade. The butchers would come in and pick out the chickens. They would buy 150 chickens at a time. I used to see truckloads of chickens coming in there and they'd kill them off over a day or two and then more truckloads would come in.

I went over there to look around. I had handled chickens before on the farm. I asked if I could get a job there. They said they had no jobs but if I wanted to pick feathers off of chickens they would give that job to me. They paid three cents a chicken. I said, 'You can make a living doing that?' They said, 'In one day, you can 'pick' up to 1000 chickens.'

I had to cut the throat of the chicken, then I would take the chicken and whack its head on a barrel. When the chicken was loose and stunned, I could get all the feathers off in two or three sweeps, pulling with my hands. It took strong fingers to hold its head, two wings and two legs in one hand. We'd whack them on the barrel, boom, boom, and pluck them bare. After plucking them, we'd throw them onto a tray on the floor and the chickens would walk around naked. They were dying, sometimes there'd be 20 of them running around. It was the filthiest, dirtiest job in the world. The chickens had lice on them and I got blood all over me. But if I did 100 an hour I made up to $3 an hour.

I'd be caked with filth after working there and I'd go to Florence's house. They usually wouldn't let me in the house. They'd say, 'Don't sit in the chair.' I said, 'I'll buy you a new sofa, what are you worried about.' I never bought them a sofa. But I made a good $30 in one day.

A married man was making $18 a week. I had tough hands. In those days my hands were like steel. I worked Wednesday to Thursday night, 24 hours, and then Saturday evening, that was the fresh kill after the Sabbath. Saturday night, chickens would be delivered to Dorchester for Sunday dinner. The Glicks had a butcher shop in Suffolk Square. They had three sons who all worked in the butcher shop. The sons used to kill chickens with me as a second job. There'd be four or five of us. Pusil Glick, Sam Glick, and Hymie Glick all stood in a row with me. As fast as a shepherd we would cut the chicken's throats, and clean them off. Yoshen Bear worked there, too. We couldn't hold a conversation with him, he was off the deep end, but he was a good chicken flicker.

Every Jewish family had chicken on Friday night. A Jewish woman would get the chicken and then singe all the tiny hairs off of it with a gas burner on the stove. You could smell the hair burning. They used to take a board and put salt on the meat and let the chicken sit for so many hours to cure it. Then they'd clean it and cook it.

The slaughterhouse had to get the chickens ready. A truck would come from Maine with 1500 chickens and the slaughterhouse would buy them. Some of them would die en route. They'd get smothered because they were packed tight in little crates. I'd go in at five o'clock on Wednesday along with a couple of the other guys. By Thursday afternoon they'd have 1000 pounds of chicken going to Dorchester. Blue Hill Avenue was the

main drag for food and restaurants of all kinds. They had giant delicatessens there. People bought wonderful salami and pastrami and all kinds of Jewish food. The slaughterhouse had a little automobile with a trunk in the back that laid open. I delivered the chickens to a line of butcher shops all the way up one side of Blue Hill Avenue and then down the other. All the chickens were in potato sacks. Each sack had a tag on it telling me which butcher ordered it. That street was nicknamed Goldman's Road.

Before I got the chicken job, an old couple we knew asked me to drive them to Bridgewater. I drove down the road from Brockton to Bridgewater and I never went over 15 miles an hour. I passed over the white line a couple of times and a cop stopped me. I went to court in Brockton and I told the judge the story. I went in with my white flannels and my white sweater with Florence. I had no money. The judge looked at me and said I was guilty of crossing the line. The fine was $25

plus costs, totaling $35. I didn't have 35 cents on me. He said, 'You're not going to leave unless you pay the fine.' There were about five or six friends with me. Everyone scratched together and we paid the fine.

I thought, 'I'm never ever going to go to court again if this is the way the system works.' I did nothing wrong. I drove a couple of friends in this old Reo that couldn't go more than 15 miles an hour. I practically begged the judge to let me off. He looked at me in my white flannels and the white shoes and sweater. He figured I was a rich kid, but he was wrong. I got the fine paid and I borrowed the money from Uncle Iz to pay my friends back.

I was delivering chickens when I got stopped on Columbia Road in Dorchester. The cop stepped out and put his hand up. I was all alone and I asked what was wrong. He said, 'You went through that intersection over 15 miles an hour. The state law is you have to slow down to 15 miles an hour at the intersections.' I told him, 'I'll lose my job, I just got this job.' I said, 'I have no money to pay a fine.' He said, 'That's just too bad.' I said, 'Will you take a chicken?' He said, 'Are you trying to bribe me?' I said, 'Yeah.' He said, 'Okay, take the chicken down to the drug store and put it into the refrigerator.' I took the chicken out of the bag, which didn't belong to me. I took it over to the druggist, told him to put it in the refrigerator and off I went. I thought, 'From now on I'm never going to pay a fine for anything.'

I used to speed like a madman in those days. Florence would ride with me. Every single time I went somewhere I'd get pinched for speeding. The cops that were on that route knew I was carrying chickens. South Boston cops and Everett cops would stop me so I started to give them a chicken and they'd leave me alone. I was speeding all the time. I got pinched every day for six months. I gave them a chicken and that was it. It was the Depression and they must of appreciated the food.

Many chickens used to come into the slaughterhouse suffocated. They used to push so many chickens into a crate. Some of them were a little sickly and they died. I would take those chickens and put them in hot water and take the feathers off. I'd clean them all off, clean the bellies out and make them ready for sale. I'd put them in the ice chest. I'd take two chickens with me on every run. Florence went with me as I went racing through the streets. I tried every road between Malden and Boston. There was always a cop who knew I was coming carrying chickens.

One Saturday night Florence and I were coming home. Her distant relative, Annie, was with us. I had already delivered the chickens so I had none with me. I was driving the Chevrolet sedan owned by the slaughterhouse. It had a box seat on the back that I'd fill with sacks of chickens when I was working. The cops spotted me on Everett Avenue. I told Florence 'I'm not going to stop. I didn't do anything wrong. I'm not speeding and I'm not going to stop.' I took off. I could see the cops were chasing us in the mirror. I got in traffic and swung off into a side street.

Annie and Florence were scared as hell. Just then a black cat jumped in front of me. I locked the brakes and I slid to a stop. Florence couldn't believe that I stopped for a black cat. She said, 'You're going to get pinched.' I said, 'I'll get pinched if I pass that black cat,' so I turned around. I spun the car right around on its haunches. I came back the other way and the cop was going by, hanging onto his car. I was going right by them in the other direction, and they never got me after that.

I PLAYED TACKLE FOR a semi-pro football team, The Malden Steamrollers. Florence would come to all the games. One time we were playing in Brockton. A player from the other team was going to jump on my back. Florence came running out from the stands and hit him with her bag and flattened him. They called her Juno after that. Juno was a cartoon in the newspaper about a tough girl. The game broke up because there were so many fights. A lot of people jumped in. Florence never wanted me to get hurt..."

JULIE HAD FUN AS A SEMI-PRO FOOTBALL PLAYER. OPPOSITE: SOME OF THE MALDEN STEAMROLLERS. JULIE STANDS IN THE BACK ROW WITH FRANK (SPEED) BRODSKY AND THE HOFFMAN BROTHERS.

CHAPTER TWO

The Road:

"**IN 1930 I WAS MAKING** $30 a week and dating Florence. I had no money left each week because I was spending it frivolously. I was working in the slaughterhouse and the owners said they were paying me too much money. I was getting more money than anyone else. I plucked so many chickens and drove to Dorchester for deliveries every week. The owners said, 'We'll pay you $25 a week. That's more than a married man makes.' I said, 'I'm getting paid by the piece, the same as these other guys.' They said no so I quit.

I figured I'd buy and sell chickens to slaughterhouses because I knew how to handle chickens from the farm. I talked to Izzy and Paul, my brothers-in-law. One was a mechanic and the other one worked for the Boston Blueprint. They weren't making much money. We decided to be partners and open a trucking company hauling chickens. I figured if I could make myself 50 cents on each chicken, I'd make a good living.

We went to Manchester, New Hampshire to a big general store. We asked them if we could get the lists of farmers that sold chickens to them. There were two or three outfits that had chicken trucks. Cohen Brothers used to pick up 1500 chickens at a time. Cohen said, 'I can't give you any names because the haulers always go directly to the farmers.' He said, 'If you want to haul freight I have 25 tons of sugar that comes in every week from Boston,

THE FOUR ORIGINAL TRUCKS USED TO HAUL FREIGHT FOR BEAVER TRANSPORTATION EXPRESS AND THEIR DRIVERS. FROM LEFT TO RIGHT: RED, JULIE, PAUL, THE GHOST from Domino and Revere Sugar.' He said he would pay 20 cents a hundred. We figured it out at 20 cents a pound and thought we could make a hell of a living hauling sugar. It was actually 20 cents per hundred pounds to Manchester, and 35 cents per hundred pounds to Lewiston, Maine.

We bought a used truck, a International eight five truck from the International Harvester Company. I had traded my car in, an old Oakland, and we got enough money together for the plates. We didn't have to make any down payment. In the Depression, the trucking companies wanted to get rid of anything they had. My partners, Florence's two older brothers, were married. They were getting $18 a week. We didn't have enough money to pay everybody the same so I took only $12 a week. They made $18 because they each had kids.

Joe Baker was an uncle to Ruthie Parkin, and he used to come down and visit. We met his two daughters, Dot and her sister. We had gone up there a couple of times to visit with them in Lewiston. Dot said, 'My father knows everybody in Lewiston and he'll get you freight to haul.' We went up to visit him. Her father said, 'I can get you freight from the woolen mills here going to Boston. You can bring the raw materials up from Boston.' That's how we started our trucking business. We called it Beaver Transportation Express. Express had a big gold X in it. The boys called it Beaver after their boy scout patrol, the most successful patrol in Boston.

The very first night I went to Lewiston, I went up alone with about 200 pounds of M. Winer groceries and probably another 100 pounds of stuff. We had an auxiliary tank installed on the truck so I wouldn't have to buy gas on the road. The Malden man who installed the tank wasn't too experienced in the transfer of fuel. He ran the line between the two gas tanks but there was no vacuum pump to suck the fuel out. I switched the valve to get the other fuel pumping and nothing happened.

I was climbing a grade into Lewiston when the engine quit. I was at a farmhouse and tried to get some help but I couldn't wake anybody up. I saw two big milk pails and brought them over. I had a wrench. I opened up the auxiliary gas tank and I let the gas fill into one pail. Then I punched a hole in the other pail, making it into a funnel, and poured the gas from the full pail to the main tank. My battery was dead because I had tried to start it after running out of gas. I got two 15-inch blocks of wood from the farm. I pushed the truck up the hill about three inches, and pushed the wood in behind the back tire. I got the truck to the top of that hill that way. Once it started to roll down the hill, I quickly climbed on the running board, jumped into the driver's seat, got it into gear and got it started as it went down the hill. I made it into Lewiston. From then on, I couldn't depend on the tanks, so anytime I got to a gas station I'd put in gas.

I delivered 300 pounds of groceries that day to the Lewiston Sanitary Market. The truck was huge and the delivery was small. I had picked up food in Malden. M. Winer's store stocked Jewish types of food like whitefish and herring. We took in about $4 on the trip.

Joe Baker helped us get some wool from the mills but it wasn't enough and we were slowly starving. Uncle Iz had lost his job and said he'd be our salesman. Uncle Iz stayed in Maine and started to get customers for us. He got the Lewiston Sanitary Market, and several shoe factories. Clark Shoe, Ault Williams, and Maine Shoe were famous shoe companies from those days. A lot of shoe companies were moving out of Lynn, Massachusetts. In Lynn the workers were getting too much money. They were paying very low prices for workers up in Lewiston and Auburn, Maine. The cities would pay the shoe factories to teach the people how to make shoes. We started to move shoe companies. We got another truck and we had all kinds of commodities moving. We hauled oil and steel, and everything our competition, Confection Spring, hauled.

When I first went into the trucking business, I took Florence and her sister Sally and her husband with me to the Cape. I wanted to show them the farm. It was a wintery night so they put a blanket over them in the back of the touring car. Florence and I were sitting up in front with the windshield protecting us, and the cops stopped me. They asked what I had in the back. I said nothing. They lifted up the blanket and the two of them were in there. The cops thought we were running booze.

Our first terminal was in Boston, we had been in business two or three years at that time. We had a secretary to answer the phone and our customers would call and say pick up so many pounds or tons of such and such. We'd load it into a small truck and bring it to the terminal, and we'd load a trailer up until it was full. Then we'd hook a tractor truck onto a trailer and head for Maine.

One day, Uncle Iz and I were waiting to see if the highway patrol were weighing the trucks going from Massachusetts to Maine. We were always getting pinched for overload. We waited in Portsmouth to hear from Tony Campbell, a driver we had hired. He was on his way south. After waiting a couple hours for our truck to come through I got discouraged and headed to Kennebunk. There was our truck, parked in Kennebunk. I said to Uncle Iz, 'What the hell is he doing here?' I climbed up onto the running board and took a look inside the truck. There he was making out with a girl.

I opened the door and I pulled him off of her. I said, 'What the hell's the matter with you? You left me sitting out there and you're screwing around here.' He said, 'I got a right to have a rest.' I said, 'You don't rest when

JULIE CLIMBING INTO ONE OF THE TRUCKS FOR THE EVENING RUN TO LEWISTON, MAINE.

you have me waiting like that.' He gave me some look and I belted him. I hit him so hard that his eye split and little balls came out. I thought I'd blinded him forever. He said, 'I'm going to sue you.' I said, 'You son of a bitch, you're not going to go to Boston in my truck. You'll walk if you're going to sue me.' I put a patch on his eye and he drove back to Boston that day.

We were driving on Newburyport Turnpike, Route 1 near Saugus, when we got stopped by some men. They wanted to know if we were in the union. We said no. They said, 'Well you ought to join.' We didn't say too much. Most truckers joined the union but we were family. We didn't need the union. When I got into Boston, another truck came through and the driver told us about a shooting. A farmer was on his way to Boston behind us. We didn't know him, but the union men tried to stop him. We were told he wouldn't stop for them, he just kept right on going. The trucker said they chased him, shot him and killed him. The next day, we saw in the paper that a farmer was shot and the police didn't know who did it. It was the union.

The farmer was carrying fruits and vegetables into Boston. They had terminals where you'd deliver vegetables and they'd sell it there. Truckloads of lettuce and cabbage came in. But he wasn't a regular truck driver. He was a farmer. That's when we decided that we would get guns. We went down and got gun permits from the police. Iva Johnson was a big store in Boston. They sold things for trucking, as well as arms. Each one of us got a gun. I bought a .38 Army Special, a big gun, for $13. Izzy got a .45 and Paul had a .32, which wasn't as big. Izzy and I were on the road all the time. Paul was mainly in Boston at our terminal. I never had to use my gun. I'd pull it out sometimes and they'd see that and they'd keep on going.

Saul went to work for us and lived in Auburn, Maine. He did the deliveries and the pickups in Lewiston with helpers. Izzy and I would pick up the freight in Boston, and I'd help load it. I drove six nights a week round trip from Boston or Malden, sometimes Malden to Lewiston and back. That's after having loaded 16 or 18 tons of freight. Every pound was stacked by me. Then I'd drive all night. Florence used to ride with me and keep me awake I was so tired. She didn't know how to drive but she used to hold the wheel while I had my foot on the gas and when a car would come she'd shake me. She'd say 'A car is coming,' and I'd open my eyes.

BEAVER TRUCKING HAD a terminal on North Street in Boston, right near Paul Revere's home on the corner of North and Commercial. A man walked up to me and looked me over. He looked like a Mafioso type. He said, 'I hear you're a good wheelman.' I said, 'What's that?' He said, 'You're a good driver. I'd like you to drive a truck for us to New York and New Jersey and back. I'll pay you good money for it.' I asked what was in it. He said, 'What do you care what's in the truck?' I asked how much I'd get. He said $75. To make $75 we'd have to haul ten tons of sugar. I agreed to do it.

They had an automatic truck, refrigerated, and there was nothing in it. An Italian fellow went with me, a tough cookie, and gave directions. I drove down to New Jersey underneath the Polaski Skyway. I went into a huge garage called McKessin and Robbins. There were all kinds of vehicles going in there picking up booze. Little vans from a flour company to big Greyhound buses, they all loaded up and brought the booze everywhere. Booze was brought in by ship into the Jersey port, and then drivers brought it to the big garage and loaded up all the different trucks. I parked their truck in the garage. They took the truck and said 'You sit here and wait.' I arrived at about 7 a.m. and waited all day. At about 5 p.m. the truck showed up loaded, with the doors sealed.

Just before we left, some men came over and talked to my guard. They brought him into an office while I waited with the truck. He came out with a shiner. I said, 'What happened?' He said, 'Well, I guess I got caught. I gave the loading guys $20 to put three extra tins on the truck for me.' They were selling Belgian alcohol for $35 a gallon, and there were five gallons in a tin. The men noticed some tins missing. They fought with him and said the next time they'd kill him. Some men gave us protection to the middle of the Holland Tunnel in case of a hijacking. They carried guns and watched the truck until we reached the end of their jurisdiction. I got to the middle of the tunnel and came out the other side without them. My guard said we had to be careful from that point on. We could get hijacked. We got back to East Boston and went into an American Express Garage. Some other men took the truck and I left.

About a week later they asked me if I wanted to make

another trip. I said it wasn't enough money. They offered me $150 for driving the truck. Once again I made the run and the same guy was my guard. The third time they didn't have the truck. It was busy. They asked me to take my truck and offered to pay me $350. Wow. I was going down with my empty GMC T42. It had a 2½ ton truck rating so it could handle five tons of weight. A four-ton truck held eight tons. I went alone into this garage. Some men took the truck from me and loaded it. At 5 p.m. they brought the truck back. It was overloaded, tins were even stacked on the tailboard. I said, 'That's too much of a load for me.' The truck had about seven tons on it, but they didn't take any tins off. They told me to take it easy and I headed out.

I was worried about being hijacked. I thought, 'I'm not going to go back the way they think.' I went up the Hudson River toward Albany. I rode down a highway called Jacob's Ladder, into Malden, then drove from Malden to Maine. I got on to Route 1. I was not far from Newburyport when my inside right tire blew out. I didn't have a spare. I was going slow but I didn't want to stop. On the south side of the Merrimack River was a big gas station, Cashman and Condens. I didn't want the people at the gas station to see the truck so I parked on the north side of the river. I jacked up the truck, put her on blocks and unhooked the inside tire. It had a bud wheel with so many studs and seemed to take forever. I took the blown out tire and I rolled it back over the bridge to the gas station. They didn't have a tire but they had a tube. I put a big patch in the tire and replaced the tube. I pumped it up and rolled the tire back over the bridge.

The truck had fallen off the jack and it was leaning against a telephone pole. It was already daylight. I had driven through the night and I was shaking in my boots. I took bricks and blocks and jacked it up a little at a time until I got it to stand up straight. I finally got the truck tire on. It took me about two hours.

I drove towards Ogunquit and wasn't going too fast on a narrow road when the tire blew again. 'Oh Jesus,' I thought, 'what am I going to do now.' I still had the outside tire. I crawled to Ogunquit in the truck so the tire wouldn't blow. I pulled over on the side a bit. I was standing looking at the tire and a state trooper came up. He was a big tall guy named Littlefield. I knew him from making the runs for Beaver Trucking. He said, 'What are you doing here?' I said, 'I got a flat tire.' I didn't know if I should run in the woods and hide or not. I could smell the booze on the truck and I didn't know if he could smell it. I said, 'I can't get a tire here in Ogunquit.' He said he'd take me back to York in his sidecar.

I was holding the big blown tire on my lap in the sidecar. Those tires weren't big enough for the load I was carrying in the truck. I had four tires in the back and two in the front. Littlefield took me to York and they didn't have a tire but they had a tube. Once again I got back to Ogunquit and I ran into a gas station with my blown out tire and made a patch. Littlefield left me at the truck and I got blocks of wood and got it all jacked up. I put the tire on the left front where it ran the coolest and had less weight. I took the front tire, which was my better tire, and put it on the inside rear. I drove very

slowly into Auburn and Lewiston. I got into Lewiston about 8 p.m. that night. I thought for sure the cops were going to grab me.

The guys I worked for met me in Lewiston. They hollered at me because I was so many hours late. I was supposed to be there in the early morning so they could unload the truck. They took me to the drop. I backed up to a barn and they unloaded all the booze. It was seven or eight tons of booze. I drove into Bangor to meet the crowd I was working for. I found them at the Bangor House and went up into their room. They gave me hell

and said they would have given me money for two spares. I couldn't afford a spare at the time. I fell asleep while I was talking to them. When I woke up they were gone. I drove home with $500. I was paid $150 more because I didn't stop in Boston, I went straight up to Lewiston. That money was for the company, I didn't get any of it. That kept Beaver Trucking alive.

My brother-in-law Izzy ran some case goods one time and he took his wife with him. He had the booze in the front of the trailer, about 200 cases of Kaister. I went up to Maine with my car and I couldn't find them. I finally found them behind a shoe factory unloading the booze. I said to Iz, 'Why did you take your wife on this goddamn Indian adventure?' Saul and Red Neirman were there helping him and we gave him hell. They unloaded the booze and we got another $300.

One time Izzy and Red Neirman were in between Auburn and Lewiston in an ice storm. They slid down a hill and into a slow-moving car in front of them. They pushed the car up over the other side of the hill. They didn't stop. It was so icy Izzy kept on going. The booze was going to Augusta and it was case goods. They got rid of the booze and brought home more money.

Red bought a Ford car and built up the back so he could put gallons of booze in the trunk. Red got arrested for rum running on his own in Connecticut. He wound up in jail for six months. I think the people who gave him the booze turned him in. They had deals with the Feds. The police had to arrest people once in a while, to show that they were catching some men. I always figured that they squealed, since it was only 75 gallons. Red still got into the military during the war. He was an MP in Paris.

Ovide Ducherme had a dairy in Lewiston and we delivered sugar to him. We brought a truck up to Maine and left it parked in the street with ten tons of sugar in it. Overnight, Ovide would pick up our truck and bring the sugar back to a secret still where he made booze. We never went to the still because that would be dangerous.

On the next trip we had 1500 gallons of booze to deliver. Ovide had a barn for storage of booze but we didn't know where he had the still. We unloaded 1000 gallons of booze in five-gallon tins into the loft of his barn. The other 500 gallons was going to Bangor. I was exhausted. I said I couldn't make it. The men I worked for said, 'You've got to. They're waiting up there for you.' I headed to Bangor with Red Neirman.

When we got close to Bangor a car flagged us down. I wouldn't stop, I thought it was somebody trying to hijack me. It was Weinstein, the bootlegger in Bangor. He said we had to go with him to the drop and unload the booze. I said, 'I don't want to know where your drop is because if you get hijacked you'll blame me.' The drop was always a place in the woods, either a barn or an underground bunker. He begged me to help. Red said, 'Let's help him.' We followed him with our truck into the woods.

We unloaded the booze and then he wanted me to take 800 gallons back to Boston. This booze had come in on a Belgian oil tanker and Weinstein couldn't sell it. When they pumped this Belgian alcohol into tanks, the tanks were saturated with oil. The booze smelled like oil, and made people sick. The bootleggers in Boston were going to replace it. We loaded the 800 gallons of tins into the truck. Red moved the truck without telling me and a whole row of tins fell over and split. Alcohol leaked all over the floor of the truck. I gave Red hell.

We got everything packed and stopped in Skowhegan at a diner on the way back. We went inside and had something to eat. When we came out we could smell the booze, everything was running through the floor. There was a state police car parked right next to us. I told Red,

'We have to get the hell out of here.' There was a pool of alcohol on the ground. We got in the truck and never stopped until we got to East Boston. We were going like hell, we figured any minute they'd stop us. We got into East Boston to the American Express garage and we got our cash.

The Feds came to my office in Boston the next day. They saw me loading the goods into Ovide's barn and said they were going to take me in. I said they couldn't take me in because no one arrested me at the time. It turned out that they didn't want me, they wanted Ovide. They knew he had a still going and he was selling booze.

There were 500 gallons of booze to move in Lewiston. Saul picked that up from the drop, and brought it to our Lewiston terminal. He drove around the city in an open truck with all the tins of booze sitting on there. They looked like containers of chemicals, luckily no one knew what was in them. When I got there that night we loaded the booze onto a regular truck and took it back to Boston.

After the delivery, I had to collect my money in East Boston at an address on Maverick Street. I rang the bell, and a guy opened the door. He looked at me and said, 'What are you doing?' I said, 'I came here to collect my money from Louie Lynch.' Lynch was a Jewish guy but changed his name. They let me in.

Inside was a speakeasy. There wasn't any dancing at that time of the day. They had a counter and guys were sitting there drinking. They had a couple of old-fashioned slot machines. There was a cop in uniform playing the slots. I was sitting there and some guy asked if I was hungry. I said yes. He said, 'Well, go in the back room, we have veal scallopini in there. A giant oven held a huge pan of veal scallopini on it. It was eight inches deep, six feet long and three feet wide. Some guy was mixing it with a big ladle. I must of eaten a gallon of that stuff. Afterwards, I sat down and waited until someone came in. A man said, 'Louie Lynch is not coming, but here's your money.' All I was interested in was the money. I figured that would be the end of rumrunning but they kept calling me.

Ships of case goods came in from Canada onto Old Orchard Beach. I pulled out on the beach waiting for a delivery and I saw a Coast Guard patrol boat coming in. I tried to pull off the beach and I sunk. I was driving my new GMC truck. It had a big orange body on it. I thought, 'They're going to see this and I'll be in trouble.' I couldn't get over the sand. I saw a cow chained up, off the edge of the beach. I ran over, took the chain off the cow and let it go free. I ran back to my truck. I wrapped the chain around my wheels. I got out of the sand and headed back to Boston. I brought the truck directly to a paint company and had it painted blue.

Another load of case goods came in by ship. The rumrunners I worked with unloaded it in Brunswick and were almost caught by customs. They left the booze and took off. The Coast Guard took the booze and put it in a warehouse. I was asked to drive a Cadillac touring car to Lewiston and then on to Brunswick. I didn't know the car was loaded with guns. Izzy, my brother-in-law,

FLORENCE
LERMAN MADE
HER OWN DRESS
FOR HER
WEDDING. THE
HONEYMOON
WAS A MERE
SEVEN HOURS.

and Louie Bear were both going to drive trucks up to Brunswick and get the load of booze back.

The government knew they were in the area. I said, 'Don't do it.' I thought they were going to get caught. Finally they decided not to hijack the booze. The bootleggers told me to take the guns and bury them somewhere. I buried them behind an A & P warehouse in Portland, Maine. A few of the guys came home in the Cadillac with me.

When we got to Biddeford Five Corners the cops stopped us. They knew I was with bootleggers. The guy next to me handed me his gun. I said, 'What the hell's the matter with you. We were told to get rid of all the guns.' There was a plastic cover in the car near my knee. I dropped the gun down there and the cops didn't see it.

The cop talked to one guy in the back. He said he was the boss. He said, 'I own a canvas business and these are my salesman.' The cop looked at me. He said, 'I know you drive a truck for Beaver. What the hell are you doing with these guys?' He said, 'Get out of here.'

I got on the road and then I saw police lights behind us. I said, 'That's them.' I flew down the highway doing 80 miles an hour. I got to Route 1 in Danvers. I was right in front of a Mobil gas station when I blew a tire. We almost went off the road. We almost got killed. I controlled the car and slid into the gas station. We changed the tire in five minutes and were gone before the police came. I was over the hill in Danvers and back in East Boston in no time. We had to rip the car apart to get the gun out.

One day, the same guard who drove with me on the first trip, said, 'Goldie, you haven't been pinched yet.' I said no. He said, 'You know, if you get pinched it will ruin your life.' I said, 'So.' He said, 'Why don't you get out?' I said, 'Will they let me out?' He said, 'I'll arrange it.' I said, 'But they owe me $1100.' He said, 'Do you want to get out or do you want the $1100? You've got a choice.' I got out of it and I never looked back. I called up Izzy and told him to come home. I said, 'Don't have anything to do with those guys.' That was the end of our rumrunning days.

I LIVED FIVE STREETS away from Florence in Malden. Florence and I used to drive down to Everett. At Christmas, Everett had the most gorgeous houses. It was a very rich area. I'd go there in this old junker of a car I had. I'd say, 'Someday I'm going to have a house like that. I'm going to have a boat and an airplane and a Cadillac.' She'd say, 'I hope so.' We had dated for three years when her father and mother thought we ought to get married. I talked it over with Florence and we decided it was time.

In the army as a kid, I remember we had a huge parade ground, where a thousand horses would run at one time. I saw a big flying boat land on the water nearby the parade ground. Bob Fog was a lieutenant in the Army Air Corps who flew the plane. I walked up to it and looked at it's shiny mahogany hull. It was a flying boat, a sikorsky. I walked around it and never gave it any more thought. I just liked the looks of it.

The day before Florence and I got married, we were driving down Squire Road and we saw a sign. It read: 'Airplane rides, 50 cents.' Florence and I took a ride together for a buck. The wind was blowing around us and it just felt interesting. We flew around Malden and Revere and came back and landed.

It was an old Spartan biplane. It had two seats in the back and one in the front. Al Lechshied was the pilot. He became one of my best friends. Florence didn't say anything but I thought she liked flying. She was happy. After we got on the ground, I told her I thought I'd like to fly. That was the beginning.

We got married on June 24, 1933. I was 20 and she was 19. The wedding was in Suffolk Square in a little community hall on Bryant Street. Florence's mother needed money to buy chickens for the wedding. We had $38 and nothing more. All the profits we made from the trucking business were put back into the business. Florence's family rented the hall for about $10 and her mother did all the cooking. I paid for the chickens, which cost me $25. That left me $13. That's all the money we had in the world.

We had dinner and dancing and flowers. Florence made her own wedding dress. She looked beautiful. She was a good kid. She didn't know what she was getting into with me. We got married that night and we went to the Hotel Mangar at North Station in Boston. The staff wouldn't let us in because we didn't have luggage. I showed them the marriage license, written in Hebrew, and they couldn't read it. I told them it was our marriage

license, They said, 'It doesn't matter what you have, you've got to have luggage.'

I had Florence sit in the lobby. I rushed to Paul's in Malden and got an empty suitcase. I rushed back to Boston to the hotel. The Mangar was brand new then. It was renamed The Madison later on. It was torn down when they built the Fleet Center. It was $3 for a night, so we had $10 left. We had a nice room, it had everything. The honeymoon was just one night, about seven hours. We skipped breakfast the next morning, and I went to the North End to the trucking terminal. Florence took the train home from North Station. I lugged freight that day and drove all night the next night.

We got a set of tableware from Uncle Max and Aunt Sarah, and a bedspread and a tablecloth. We didn't get much more than that, two or three trinkets. My father gave us a banjo clock for a wedding gift. It was worth about $3. We were given sheets, pillowcases, and silverware, not silver but tinware. We had $10 and our presents between us and the rest of the world. We bought a bedroom set and had it sent to her mother's house.

Pretty soon, we had an apartment on the same street, Hazelwood Street. We rented it for $18 a month. We bought some more furniture and we made payments on it. Florence stayed in the apartment all by herself while I drove the truck.

WE WERE STRUGGLING. We made a living in the trucking business but that was all, we never made any real money. When I decided I wanted to learn how to fly, my brothers-in-law Izzy and Paul were also interested. I thought that this was the way for this trucking company to provide air freight. We had no money so we bought flying lessons 15 minutes at a time. Izzy bought enough time to get his solo. I'd stop at an airport with the truck and I'd ask them if they had anything for sale. If they did I'd ask for a demonstration ride and then I'd buy 15 minutes. It cost $4 or $5 at that time. I started building time. Out of the five hours and 20 minutes I flew in a year, I paid for maybe three hours.

Ralph Ames was an instructor for Johnnie Shobe's Flying Service at Boston Airport. He told me I was ready to solo. I bought fifteen minutes and then I flew my first solo. If the engine would have burped, I would have been dead. I knew nothing. People normally needed between 8-12 hours of paid solo time. The free time didn't count. Nowadays the average kid takes about 15-25 hours of flight time to solo. I loved it. It was what I wanted to do, I could see beyond into the sky. I wanted to be in the air.

I kept building and buying time. My friend, Nat Trager, purchased an Arrow Sport for $2250 in 1937. The plane was built in 1927. We started to fly together. We eventually decided to buy a new airplane and we traded that for the an Aeronca. It was the first Aeronca Chief built with a speed wing. We formed a flying club with Izzy, Paul, Nat and myself. We taught Red Neirman and Dave Silver to fly, and then decided to instruct people in order to pay for the airplane. We built enough time and once we passed the course, we became flight instructors.

I paid $600 to learn aerobatics, a secondary course at Muller Field. Riverside Flying School was a part of Muller Field. The school belonged to three or four guys, among them Charlie Hamilton, Jack Phillips, and Ronnie White. They owned another field south of Boston called Bayside. I got my secondary rating with Lee Hipson, a famous pilot in the area. Hipson was an inspector for the government. I got an instructors rating and I paid for it. I flew 30 hours and then I taught aerobatics after that.

It took a month to teach all the different aerobatics. Students would come for five hours of lessons a day. Every day I had five students from MIT or Tufts. We'd start out in the morning at 7 a.m. and finish at noon. Five days a week. I had a business card with everything typed on it and I followed the routine.

I was teaching flying and driving trucks at the same time. I drove nights and in the morning I'd have a class to teach. When the war started I had a class from Tufts and when I graduated that class I had a class from MIT. I taught five students at a time in a Waco UPF-7. All my aerobatics students had already gotten a private pilot license. They had 35 hours each in Aeroncas or Cubs.

There was a lot of hard driving in the winter storms. Around 1934, I was on the road with one of our drivers, Sully. I had a straight truck loaded with heavy freight. I came up over a hill and there was a trailer jackknifed in front of me. I couldn't stop. I spun around and got one wheel in a ditch and one on the pavement. I went around the trailer and got back on the pavement. The truck just laid over and slid on its side in the middle of the road.

Sully fell in behind me behind the seat and I didn't know it. I stepped on him when I climbed out of the truck. I was looking for him when he started yelling inside. I climbed up to get back on the side of the truck. I got hold of the exhaust and singed my hands. They were cherry red. I had calluses that were a quarter inch thick and it just burned the calluses down. My hands were fine. We got Sully out of the truck and he was okay. I got a wrecker to pull our truck onto its wheels. My truck had a burnt side, and the fender was bent up. I drove it home to Boston.

In 1937, we ran into problems delivering freight to companies involved with the union. The businesses wouldn't want to accept freight from us. They boycotted us at times because we weren't union members. In order to get our freight delivered, we finally gave in and joined against our will.

I went to the Cleveland Air Show in 1937 with my brother-in-law, Paul, and a couple other guys. We rented a four-seater airplane and flew out there to watch the events. I saw things I had never seen before in aviation. I saw the formations of Air Force airplanes and Navy airplanes and the Cleveland Air Races. They had the V15, the biggest airplane in the world. Its engines weren't powerful enough so they upgraded it and built the V17.

IN 1938, A TRAIN RAN from North Station in Boston, to North Conway, New Hampshire, to Mount Cranmore. We'd go on the ski train, it didn't cost much, and take our skis with us. Everybody was singing and polishing their skis and they'd get there and get up the mountain, and on the way home there'd be stretchers and broken legs. There was a rope that towed you the first leg, then they had a sledge that took you to the top. It had nine seats and they pulled it to the top. None of us knew how to ski.

My brother-in-law, Pete, made me a set of skis with no iron edges. I went to the top and I started down. By the time I got to the bottom, after falling three or four times, I broke one ski in half and they had ammonia pills under my nose to bring me to. There was a big rock at the bottom and I'd head for the rock every time.

I learned to ski a little bit but I didn't have time for it. There was a girl who skied down beside me, and she didn't know how to ski either. She hit a tree with a leg on each side and broke her thigh. They took her down on a stretcher, and she was in agony. Skiing was new then.

I HAD THE CAR PARKED on Hazelwood Street in front of our place. I had a gun in the glove compartment. Someone broke into the car and stole the gun. Florence came home one day and everything was tossed around and scattered all over the floor, including pictures of ours. She said it must of been someone we knew because whoever broke in was looking at our pictures. Florence was scared by herself. There was nothing to steal, we had nothing. When I saw what had happened I gave her a gun, a Colt 25 Automatic. I taught her how to shoot it. It was right before World War Two.

It was dark the next night when I got home. I got in about four in the morning which was earlier than usual. Normally I'd get in about six. It was pitch black and I broke my way through the door. All of a sudden I saw something moving. I switched on the light and there she was sitting in the bed with the gun pointed at me. I said, 'Put it away, you're aiming at the wrong guy.' She had the gun but she never dug it out again. She had her mother's dog with her, and it didn't make a sound, it just sat there on the floor. We lived there until we moved to Clark Street in Chelsea, in 1939.

WE WERE HEADING to Lewiston one night. My trailer was loaded with several tons of hydrogen gas in cylinders. My brother-in law Izzy followed me with the semi-trailer. It was one of the coldest winters we'd had in ages, it was about 20 below zero outside. I'd drive 300 miles a night, 150 miles each way. In South Portland, there's an overpass over the railroad yards before turning into the city. At the railroad station we usually stopped for a sandwich, ham and eggs, or coffee.

We climbed the grade to the overpass at 11 p.m. As I turned onto the overpass a train went by underneath. Those days they had steam trains that burned coal and it made a lot of smoke. The smoke covered the whole area. It just hung there in a frozen mass because it was so cold. I hit the brakes. We had installed the first air brakes ever on this truck, Westinghouse air brakes. We didn't know that you had to blow the air brake tanks every so often. A diaphragm, or holding tank, was attached to each wheel and the water that builds up from moisture in the air had settled in there and it froze. I had about 10 percent of my braking power. Even though I was going slow there was no energy to stop the heavy truck, loaded with hydrogen cylinders.

I wasn't doing more than 20 and I slowed down a bit, but the truck didn't stop. I saw red lights just to the left of my nose in the fog and smoke, so I veered to the right. I hit something in front of me with no lights. It was a car blocking the road. I hit the air hard, and I had no brakes. I turned hard right figuring I would stop against the drifts but the kinetic energy kept me going. I took the car, the truck and myself and went off the bridge down into the freight yard below.

As I plummeted off the top of the bridge, the car was rolling upside down. I landed on top of it. The truck went straight down and then the force of the crash pushed the car out in the snow. The fall was about thirty feet, and I thought I had bought the farm. I bounced around a little bit. I was jolted when I hit. I got out of the truck and I heard a gurgling noise. I thought there were people in the car. I had no flashlight, and no lights on the truck. I was fishing in the car when I heard somebody yell from up on the bridge. He said, 'Hey down there, are you all right?' I said, 'I'm okay, but these people must be close to dead.' He said, 'There was nobody in the car, it was my car.' The whole weight of the world came off me then because I thought I killed a bunch of people.

I wasn't there four, five minutes and along came Izzy in the other truck. He stopped up there with his booster brakes. He looked down, saw the truck and jumped off that bridge right into the snow. He hit a cable on the way down, and landed on his back. The ground shook. I thought he killed himself. He figured I was dead. He saw me, picked himself up and came running over. I weighed 220 pounds. He picked me up and carried me around. He said, 'We can buy new trucks as long as you're okay, you okay, you okay?' I said, 'I'm okay, Iz. You got hurt more than I did.'

My truck was in the shape of an inverted v. The rear of the truck was down and the front of the truck was down. We worked all night to get that mess cleared up. A man had been at the bottom of the causeway who had just come out of the freight yard. Another car came off of the causeway, hit his car and kept on going. This had knocked his hubcap off. The hubcap rolled up the causeway. He pulled up on the bridge, parked his car and was looking for the hubcap when an old couple in a coupe came down and they ran into him. The lights I saw on the left was the old couple still sitting in there.

By instinct when I saw their lights I turned right. I avoided hitting the car that was running, but hit the parked car and went off the bridge.

Saul was called and he came down with a loaded tractor trailer from Lewiston. We worked all night to pull out my big rig, with ropes and chains. The whole front of the tractor was crushed. We unlocked the fifth wheel, where the trailer is attached to the tractor, pulled the tractor out from underneath and put Saul's tractor under my trailer. The trailer had beams in the sides that were built like railroad cars. There wasn't a bend in it. If one of the hydrogen bottles had exploded I'd of been killed instantly. The freight was solid inside. Saul hooked onto it and he went on to Lewiston to deliver the hydrogen. We called Auto Car and they came in with a big wrecker for the tractor.

THE TRUCK BELOW CARRIED JULIE AND BOTTLES OF HYDROGEN GAS WHEN IT WENT OFF A BRIDGE INTO A RAILROAD YARD.

There were 36 bolts that held the fifth wheel, the pivoting wheel, onto the frame of my tractor. Thirty-five of the bolts broke and one held. If that 36th bolt had broken, the fifth wheel would have cut me in half. Everything was intact except all the bolts were sheared. We put some temporary bolts to hold it in place. The wrecker picked up the front of my truck and towed me into Boston. That fifth wheel would have crushed me all together. Florence said that my mother was watching me from Heaven. We got the truck to Boston and they dropped the trailer off at Sullivan Square. The tractor went to Brookline to the Auto Car company. They rebuilt that same truck, and made it look like new. We got the Boston tractor to haul the trailer in and delivered Saul's freight the following day.

IN 1938, PRATT & WHITNEY sponsored the First New England Air Tour. It went from Boston to Hartford, Connecticut to Springfield, Massachusetts to Montpelier, Vermont to Concord, New Hampshire to Augusta, Maine, to Rhode Island. There were about 30 airplanes. We didn't fly in formation but we tagged each other down. The first stop was Hartford, Connecticutt. Pratt and Whitney had a little airport, and a big factory there. We watched them build engines and that sort of thing. From there we took off and went to Barnes Field in Springfield. Barnes was just being built. It was a one-strip field. The town fathers hosted us in Springfield, and in Montpelier and Concord. When we got to Augusta we stayed overnight. Governor Grant of Maine was there. The state police from Maine backed the event. They often backed similar events because many of the policemen were pilots. We made the whole tour, with different pilots from all over New England. A magazine called *The Yankee Flyer* sponsored it.

On the way to Montpelier, Vermont, I had a forced landing in White River Junction when my engine quit. I circled down and landed. I got gasoline and took off again. When I got to Montpelier, the engine quit again. After the tour had ended, I talked to Charlie Hamilton, one of the owners of Muller Field. Charlie said, 'Don't you use carburetor heat?' He explained what it was. A venturi inside the engine compressed the air and built up oxygen in order to burn the gas. Before takeoff, the venturi was heated with carburetor heat and then the engines were all set. If the venturi wasn't heated and allowed to freeze, the engine would die. It was a brand new plane and should have been equipped with it. Charlie built me a carburetor heater for the plane.

When Charles Lindbergh took off from San Diego in 1927, he had trouble. He headed for St. Louis to show the business people their investment. It wasn't long before his plane, *The Spirit of St. Louis,* had carburetor ice problems. He figured out the problem and had a carburetor heater built and shipped to New York. Lindbergh went on east and took off from New York to France. Six men had died trying to fly across the ocean. They didn't have carburetor heat. Lindbergh had one built and when the engine got rough, he put on the heat. He was a pretty smart pilot.

In 1932, Johnnie Polando and Russell Boardmen flew a historic flight of their own from New York to Turkey. Boardman was the commander of the airplane and Polando was the copilot and mechanic. They flew a single-engine Bellanca, loaded with fuel. They flew until they had 15 minutes of gas left and then landed in a hayfield. They had the longest record of flight in the world and were heroes when they came home. They were given the distinguished English Flying Cross. Hyannis built an airport named after Boardman and Polando. Russell Boardman was killed in a race across the country. But Johnnie Polando flew most of his life around here. He was a captain in the air corps.

In 1939, I did an airshow of aerobatics with other pilots at the Plymouth Airport which was new. We stayed overnight in Chatham at the Chatham Barns Inn. During that night a hurricane hit us. We put all the planes on their noses in the little hanger in Chatham and they survived all the wind. The next afternoon we performed at Chatham Airport, built by Wilfred Berube, Sr. all by himself. Johnnie Polando did an old woman's act that day. He put on women's clothes and climbed into an idling plane and it took off. As he flew by the audience, he screamed, 'Help! Help!' He was funny. Eventually we got him down. I did aerobatics that day.

A YOUNG KID AT Muller Field didn't know much about airplanes. He was a line boy. He would spin the prop and tie the airplanes down. The airplane was tied down at three points, each wing tip and the tail. The

ABOVE: JULIE'S LIST OF AEROBATICS TYPED ON THE BACK OF A BEAVER BUSINESS CARD.

kid was sent out to start the engine and let it warm up. The owner had called and said he was coming. Some fellows when they killed the engine, would shut off the fuel and then push the throttle in, supposedly to clean out the carburetor. The throttle was left wide open.

This kid put on the ignition switch, spun the prop, and the plane started right up. It started just as I was walking out with a student. The kid got away from the prop luckily and the airplane jumped forward. The left rope and the tail rope tore and the plane started spinning around. My student saw the plane and fell down in the mud. The wheels passed on each side of him, missing him by inches. The prop just barely missed his head.

As the plane kept moving, the last rope broke. I was running away but the plane wing knocked my hat off. Pilotless, it went out to the middle of the field. Then a gust came and the plane took off, just like someone was in it. It climbed up over the quarry. The gust lifted it and it came down right at the entrance of our seaplane base at the edge of the highway on the sidewalk. It blew up. I had a white Chrysler and I ran to my car and the student covered with mud said, 'I want to come too,' and we drove over the hill and went to the accident scene.

It crashed onto the edge of International Highway. The Boston American newspaper reported on it. 'Missing cars in traffic by inches as it crashed, the plane burst into flames as traffic came to a brake-screeching stop.' I was shaking my head saying, 'Goddamn it.' Buddy Warwick was standing beside me. He said, 'Whose airplane is it?' I said, 'It's yours.' He was the son of the owner of Warwick Bus Company. He said, 'Oh, geez, I only told him to warm it up.' I said, 'That will teach you to warm up your own airplane.'

In 1938, I was friends with a young pilot, Meyer Goldman. Meyer had gotten a private license. He took his friend, Bennie Greenspan, for a ride in the airplane. He was in an awful hurry. As I saw him taxiing by I put my hand up and stopped him. I said, 'What's your hurry?' He said, 'We've got only a few minutes and I want to take him. He's never been on an airplane before.' It was against the law to carry a passenger with a private license and have dual controls hooked up. I reached in and I pulled the pedals out which were on quick disconnect. I pulled out the wheel and put both items in the baggage compartment. Meyer taxied down the runway that went downhill, and took up over the hill, over the wires going towards the city in a southwesterly direction.

I was talking to Mush Cataldo, a boy who had a Curtiss Robin airplane, and Tom Brawdy, Sr. I saw Mush Cataldo's eyes open up. I look around in the sky and I didn't see our airplane. I looked toward the marsh and I saw the tail sticking out of the marsh. He had made a climbing turn at a very slow speed and the plane spun in, it was going too slow.

Benny Greenspan was laying in the wreckage when I got to the marsh. I could see Meyer was dead. They were pulling Benny out of the marsh and I could see him putting his hand up. He was putting his finger in his eye saying, 'Meyer killed me, Meyer killed me.' His eye was hanging out and he was trying to poke it back in.

MULLER FIELD IN 1939 WAS A BUSY PLACE UNTIL THE SECOND WORLD WAR. THE U. S. WAR DEPARTMENT CLOSED ALL AIRPORTS WITHIN FIFTY MILES OF THE COAST.

FLORENCE
CATES, JACK
PHILLIPS AND
MARION
MCINTIRE AT
MULLER FIELD,
1939, IN FRONT
OF A WACO
UPF-7.

I took his hand away and we got him onto the stretcher. Benny and Meyer were sent in ambulances to Massachusetts General Hospital in Boston.

I drove to the Goldman's delicatessen in Malden and told Meyer's father about the crash. He said, 'How bad is he hurt?' We were on our way to Mass General. I said, 'Well, he's hurt bad.' When we reached Everett Station, his father said, 'I don't believe he's hurt, I think he's dead.' He asked, 'Is he dead?' So I told him the truth. He opened the door and tried to jump out of my car. I grabbed him and pulled him in.

I figured Benny Greenspan would be blind forever. But they drafted him into the Army, in World War Two. He had a plate in his head and was blind in one eye and they drafted him. There was quite a to do about flying with the family after that. All the wives said that all the guys from Malden better stop flying with their families. Izzy and Paul stopped flying, everybody stopped except Trager and myself.

Larry Hanscom bought a course at Muller Field for $600. He was a newspaper photographer for the *Worcester Telegram*. He said he was going to join the Canadian Air Force and go to war with them. About a week after he bought his course, I bought my course. I had to do aerobatics with an instructor and then I had to go practice by myself. They had a new Waco and another plane, a Fleet. Larry came down one day and the Waco wasn't ready to use. He and his friend, Dr. Cabone, went out for a ride. He was doing aerobatics over the south end of the field. He was doing loops and on top of one of the loops the engine quit. They said he fell out of an inverted spin and went in, and killed the two of them.

I went to Lexington for the funeral. I was sitting in the church and I saw this tall fair gentleman. It was Governor Saltonstall, he was the governor of Massachusetts. He said, 'I think I have to do something for Larry.' He knew him because of the *Worcester Telegram*. They were just building the strip in Bedford then and that strip became Hanscom Field.

Larry's girlfriend, Florence Cates, was a pilot. She belonged to the Luscomb Flying Club at Revere. After Larry died, they would not let her fly. I saw her at Muller Field sitting on the bench and said, 'What are you doing?' She said, 'I want to fly and no one will let me.' I said, 'Come on, I'll go with you.' I was an instructor by then. We got into a Luscombe plane and flew around and she did a nice job. She came in and landed. I said to her, 'Go.' She said, 'No kidding?' I told her to go. She had soloed before but they were afraid she would kill herself.

IN 1941, I WAS TEACHING in Revere when a flight school opened in Beverly. Nat Trager was teaching at that school and said they had about 80 students. They needed instructors badly over there. He said if I went over there they would treat me well. Warren Frothingham owned the flight school. He called me and asked me if I'd come over and teach. They hadn't started building the runway yet. It was a grass field when we started flying out of it, then the military came in and started to build one runway. It had ten hangers and an

FLORENCE
CATES AND
LARRY HANSCOM
WERE DATING AT
THE TIME
OF HIS DEATH.
HANSCOM FIELD
WAS NAMED IN
HIS HONOR.

administration building. I taught students with Nat Trager and Dr. Nangle and Howard Dutton. Dr. Nangle was a flight surgeon and Howard Dutton had a little airport in Haverhill after the war. Both were instructors on the Civilian Pilot Training Program.

We had about ten instructors and so many kids to teach. They needed two flight examiners to give them all private licenses when they passed their flight tests. Royce Kunze was the senior CAA inspector. He made me a flight examiner. I was one of the first flight examiners in the country for the Civil Aeronautics Association, now called the Federal Aviation Administration. I flew with him on aerobatics and on regular primary. Howard Dutton was also made a flight examiner.

I kept on flying and driving a truck. I hired someone to cover me some nights because I was so tired. I convinced my brothers-in-law we ought to go into the air express business. I bought this old Curtis Sedan airplane sitting at Logan at Intercity Aviation. Its owners were Nancy and Bob Love. Nancy Love eventually became famous in the war and is featured in the Smithsonian Museum. The plane had a 49 ft. wing span with a J67 engine. It had 250 horsepower and it could carry anything you could stuff in it. You could get 15 people in it but it was only meant to carry four.

I decided I'd have the plane fixed up because it wasn't in good shape. Sam Bear had bought it from Arthur Farnsworth. He was the husband of the actress, Betty Davis. I was told that Farnsworth got drunk and fell down the stairs one night. He broke his neck and died. The airplane was sitting and waiting for somebody to take it off their hands. Sam Bear bought it, but he couldn't fly the airplane and it sat there. I heard about it so I called him. At first he wanted $1500 but I got him down to $1250. I wanted to get the airplane licensed. I took it to Intercity Aviation at Boston Airport. They found a crack in the nose cone. It was leaking oil, and I had to charge the battery each time to start the engine. It needed radios, and it didn't have a regulator or a generator. The 49 foot wingspan was all fabric and some of it was bad.

Someone at the Wright Cyclone Factory in New Jersey said it would cost me about $3000 for a new nose cone. I decided I didn't want the airplane so I called Sam from a telephone booth. He said I had to take it off his hands. He asked 'How much will you offer?' I said $750. He said okay. This was my second airplane for our little company of Izzy and Paul and myself. I said I'd start the air express.

I heard there was an engine at the Ringe Tech Trade School in Boston. I talked to the school instructor and asked if he had a nose cone on an engine the students were working on in school. He had one in good condition. I said, 'Were you ever going to use that engine for flight?' He said, 'No, it was given to us by the Army Air Corps for the school.' I asked if I could have it, and he said, 'No, we've got to have a nose cone.' I said, 'I'll give you a good cracked one, you can't even tell it's cracked. You just take it apart and put it together over and over.'

The school instructor was a young guy. He needed an A40 continental engine for an ice boat. Ice boats were triangular boats with three skates and a sail built just for fun. If the wind died, a motor was used to get it going again, or to turn the boat in the right direction. I made an ice boat at the farm. There was a lake nearby and I tried my boat out but I couldn't control it very well without a motor. I got across the lake and then had to push it back. I went to Wiggins Flying Service and they had parts in a huge box. The chief mechanic said, 'I will give you enough parts to make an engine.'

It was obsolete stuff and most of it was cracked. I fished around in there with one of the chief mechanics and got all the parts to make a single-meg engine. The chief mechanic at Wiggins said he had a Waco with a bad rear case, also a $4000 item. He knew that the school had a good rear case and he wanted it. I said, 'What will you do for me, I need a generator and a regulator.' He said, 'If you can get us a rear case we'll install a new generator and regulator on your airplane free.'

I brought the engine parts over to the instructor at Ringe. I exchanged that box of parts for the nose cone of that good engine. I said I'd like to trade rear cases as well, and he said okay. I exchanged the old rear case for the new one and brought it to Wiggins Airways. The chief mechanic installed a generator and a regulator in my plane. I bought an RCA ABC one-channel radio for the transmitter and an ABC 111 for the receiver. Wiggins installed it for me. He put an antenna on that I cranked with a wheel. I put it out on a little windsock and cranked it out 100 feet to the wave length of that radio. I could call up to 1000 miles away. The radio had only 10 watts but it was powerful. I had a two-way radio, instruments, and electricity. I started to fly nights and learned how to fly instruments.

I was flying charters for my own business, Beaver Air Express. I was getting $10 for each package that was to go to Maine and it was costing me $50 to fly them. I lost money on every flight but I was flying. Sometimes I'd take one package per flight. Whatever I had I flew, I just wanted to fly. I had a calendar made to advertise Beaver Air Express. I used the plane for training people to fly, flying charter, doing whatever I could.

AT THE START of the war, the government had called the National Guard into active duty. Shortly after, the Massachusetts State Guard was formed. I met Earl Boardman, the brother of Russell Boardman. Earl had financed all of Russell's airplanes. He was a wonderful pilot and was very friendly to Florence and myself. Earl became the commander of the Massachusetts State Guard and commissioned me to 2nd Lieutenant in 1940. My cousin, Sonny Goldman, and I joined the State Guard with my brothers-in-law, Izzy and Paul. We did local missions and whatever they asked us to do.

When Sonny was young, a kid smashed a glass with a bat. A piece of glass went flying and Sonny was blinded in one eye. He was about eight years old. It didn't affect his appearance at all. He did everything with the one eye that every other kid would do with two. He played football on the Malden High School team, he drove, and it was alright. Five months after joining the State Guard, we all got medicals. They discharged Sonny for blindness. Soon after, he married his girlfriend, Marion.

Every man had to carry a draft card. I was undraftable since my business drove war supplies. We were picking up gun barrels from Maine and doing wartime work. I was also married with two kids. Sonny thought he wasn't draftable because he had been discharged from the Guard. Two months after he got married, he was drafted into the U. S. Army. He was in Providence for a while and then they sent him to England.

FLORENCE WENT ALONG with me with everything. If I made I made, and if I lost, I lost. She was a great lady. Anything I did was okay with her. She wanted to work. She worked at Malden Mills Woolen Mill, and at Woolworth's, the five-and-ten. I told her I didn't want her to work. I said, 'If I can't support you I don't want to be married.' She never worked after that.

Florence was alone a lot of the time. She used to play Mah-Jongg, a Chinese game, with her girlfriends. She had a club that she was involved in all the time, and she kept pretty busy in the house. Florence and I wanted the same things. We knew we wanted children. She wanted to adopt a child but I wouldn't go for it. I was an orphan and I saw how people used to look us over like little statues. 'This one looks good, that one looks good.' We tried for nine years before she became pregnant.

When Florence got pregnant, we were living in Chelsea. She was six months along when she had pains. I called the doctor's office and they said her doctor was up at the Mass Women's Hospital on the hill there in Boston. I was driving through Boston, and I didn't know where the hospital was so I stopped a woman on the corner. She climbed in. She thought I was picking her up. Florence was on the back seat. I said, 'Get out, get out, my wife is here, get out!' She said, 'The Mass Women's Hospital is up there.' Off I went up the hill. I got there and brought Florence in. The doctor came out eventually. He said, 'I don't think anything is going to happen, go home.'

I wasn't home but a couple of hours sleeping, and the phone rang. The doctor said, 'You're the father of two little girls.' I said, 'No kidding.' He said, 'You better come up here because I don't think they're going to live.' I was in Chelsea and I drove there in 15 minutes. It was St. Patrick's Day, March 17th, 4 a.m., and it was snowing. Florence hadn't seen the girls yet because she was still sleeping. I ran to maternity and they were lying in a little crib. One of them weighed one pound 14 ounces, the other weighed two pounds one ounce. Their eyes were closed. They were as tiny as two little mice with hair on their face and perfect little fingers. Each hand wasn't bigger than my whole fingernail. I said, 'Can't we do something?' He said, 'They may be able to help them at the Children's Hospital.' I said, 'Where's that?' He said, 'At the bottom of the hill.' I headed down there.

I went down and the Children's Hospital section was closed. They opened at eight o'clock. It was snowing outside. I waited for an hour until the door opened. I went in and met the head nurse. I told her my story. I pleaded with her to help the girls. She said, 'We can't take two babies from one family. We only have room for ten babies total, and we already have eight. We can only take one.' I looked at her and said, 'Would you be kind enough to tell me which one you want to die?' She paused and said, 'Okay, we'll take the both of them.'

I went up the hill with an intern and an incubator. We went into maternity and the intern put them into the incubator. It had lights and heat and cushions. The two little girls were lying in there, and there was room for twenty of them. We went back down the hill, and they put them in a secluded area, not near the other eight babies. I could see their hearts beating very fast and I called the nurse. I said, 'Something's happening to my kids.' She said, 'Oh no, that's what happens when they're that weak. They are trying to survive.'

I left after a while to see Florence. When she woke up, I told her the story. Marlene Iris was named after my mother, Molly, and Iris for St. Patrick's Day. Myrna Gale was named in memory of Meyer Goldman, my friend who had died in the plane crash. Her middle name was Gale because the wind was blowing up a storm that day.

The head nurse told me the babies needed mother's milk. Florence pumped six or seven bottles of milk a day for months. Every day we'd drive the milk to the hospital and see the babies. The twins were so small they wouldn't drink more than a thimblefull at a time. After four months we brought Marlene home. Florence had a mothering instinct and knew what to do. Two weeks later, I brought home Myrna and now Florence really had her hands full. They were crying babies. They were squawking and screeching and she was trying to feed them with the droppers. They were still little things, only five pounds each.

I went back up to the hospital, and saw the head nurse. I said, 'How much money do we owe you?' She said, 'All you owe is $32. We've been giving you credit, for every ounce of the milk you've been bringing in. It virtually paid the bill.' What a great lady. Florence worked her butt off trying to grow those kids."

RIVERSIDE PILOTS ASS'N.
DINNER AT MEDFORD MASS.
MARCH 29, 1939.

CHAPTER THREE

Flying Time:

"IN DECEMBER OF 1941, I was asleep at
the National Guard building around 7 a.m. when
someone woke me. The Japanese had bombed Pearl
Harbor. I couldn't believe that they'd do such a thing,
but it was done. After that, everybody I knew was
signing up for the draft and going into the military.

I was flying my Curtis Sedan for the State. The day
after the bombing of Pearl Harbor, an antiaircraft battery
came into Logan from Pennsylvania. Vic, Florence's
brother had been drafted into the army. He was down in
Camp Edwards and his unit was headed to the Fiji Islands.
He called me and told me he didn't want to go. He asked
if I could get him out of it. I said I could talk to Captain

Pompernick. He was the head of the antiaircraft outfit
that just came in to guard Boston. I said I'd see if I could
get him into his outfit.

I was flying Pompernick locally, as well as spotting
for places to put in gun placements. Pompernick said he
couldn't ask for Vic to join his outfit. He told me to have
Vic ask for a transfer. If Pompernick saw it go across his
desk, he said he would remember the name. Vic applied
for a transfer and his commanding officer agreed to it.

First thing I knew, Vic was at Boston Airport,
sleeping in a pup tent in the mud beside an antiaircraft
gun. They gave him an I. Q. test. Vic was pretty bright—
he had an I.Q. of 131. He was taken out in the field and

MEMBERS OF
THE RIVERSIDE
PILOTS
ASSOCIATION OF
REVERE BEFORE
THE WAR. JULIE
IS SEATED IN THE
CENTER.

put in charge of supply. He was there about two weeks when he called Pompernick. He said he didn't want to be there, he wanted to be out in the field with the guys. Pompernick called me up. He said, 'He's crazy, they can have him commissioned in a year.' He said, 'Well, if that's what he wants.' He put him back out in the mud with the other guys.

IT WAS FEBRUARY, 1942. I was married and had two children at the time. The kids were about eleven months old and we were living in Chelsea. We had a war business which meant that we were draft exempt. I made up my mind I wanted to be in the Air Corps. I talked to Florence at length about it. I said that I wanted to go. She said, 'The only thing that I want you to do is tell my mother that you were drafted. You didn't go on your own.' I said okay.

I took a military medical and found out that my nose had a deviated septum. I had been punched in the nose years before, and the doctor said I couldn't pass the medical. I asked what to do. The doctor said, 'Get an operation, and furthermore you have to take off 20 pounds.' I talked to Florence about it. We decided I would eat nothing but vegetables and meat. No bread, no cake, no candy, nothing with fat. In 30 days I took off the 20 pounds.

A Dr. Wolfson was on Bay Street, in Boston, near Fenway Park. I went in to see him in truck driver clothes. I told him I had a deviated septum and needed an operation. I asked how much it would cost. He said,

'Well, since you want to get in the services, I'll do it for $50.' The next day I went to a small local hospital and the doctor operated. I had a hemorrhage and my eyes turned black. Lillian heard that I was there. She came in and was my nurse for the night. She said I was going crazy from pain. The next day they let me go home.

The Army Air Corps was accepting applicants for pilots at Little Logan Field in Baltimore. Al Lechshied was the pilot who took Florence and me on our first flight. He and I were good friends and drove to Baltimore together. We both stayed a couple days and took flight tests in PT-26s. We passed the flight tests but there were some questions about Al's background. The officers started to interrogate him. Al had flown in World War One for America but his family was German, of Bavarian descent. The minute they started talking to him he got defensive. They called him all kinds of names and said, 'Get the hell out of here.' He said, 'We're going home.'

I left with Al, although they wanted me to stay. They had asked me to be a civilian pilot for them. I had lied about going to high school and was worried that they would find out and never commission me into the Air Corps. Before we left, I asked when I'd get my commission. They said, 'Well, you'll get it in time.' I said, 'When you give me a commission, I'll be back.' I left with Al and I went back to instructing. Al went to work as a test pilot for Curtiss Airplane Company in Buffalo, New York. From there he went to work for Northeast Airlines in Buffalo as a check pilot.

Because of the war the government said planes

couldn't fly within 50 miles of the coast. Beverly Flight School moved to Claremont, New Hampshire. I was an examiner there and gave flight training. I stayed in New Hampshire during the week, and flew home on the weekends. The closest airport near home was in Grafton, New Hampshire. Riverside Flying School at Muller Field was closed. Bayside Flying School, their second operation, became part of the shipyard south of Boston.

NAT TRAGER WAS TALKING about opening up an airport operation in Huntingburg, Indiana. He saw an advertisement which said they were looking for a fixed base operator. He asked me to be his chief pilot. I talked it over with Florence and said I'd see what he had to offer. He called the people on the phone and arranged it. They offered him all kinds of incentives. He bought two airplanes, and we left Claremont, New Hampshire.

My brothers-in-law were running the trucking company and I went off payroll. Beaver Air depended on me to fly the airplane so that stopped. I sold my Curtis Sedan to a guy in Saginaw, Michigan. I wanted to get the boys back the money they invested, and make sure they didn't lose out. We bought the airplane primarily for my own flying. I paid $750 for it and I got $2500 after using it for all those hours.

R.J. Paul sent me a deposit for my Curtis Sedan. Trager and I flew out to Michigan to drop it off. We made wonderful time. We made one stop in Buffalo and then flew to Michigan. Paul couldn't fly the airplane—he

was so tall he couldn't get the yoke over his knees. We went to the next airport at Bay City. We found a mechanic who could saw the legs off the seat and lower it. Then I found an instructor in Saginaw to teach him how to fly the Curtis. I checked the instructor out in the plane. Paul gave me the rest of the money and said he'd get more lessons. We took the train home.

Trager bought two Piper Cubs for his operation in Indiana. There were no radios installed in the planes but I kept a little radio in my pocket. We took off for Indiana in a snowstorm. Trager had a mechanic with him and I had the baggage. I asked, 'Trager, did you check the weather in Albany?' He said, 'I checked it. It's clear.' By

JULIE'S SIBLINGS. LILLIAN BECAME A NURSE AND SAUL STARTED TWO SUB SHOPS IN CALIFORNIA.

the time we got to Manchester, Vermont, we couldn't see a thing. I was up high over the mountains. Trager was down low in the valley and I was watching him down there. All of a sudden, he spiraled down and landed on a farm field covered with snow. I spiraled down and saw Trager and the mechanic were waving, telling me to come in and land. I figured there had to be something seriously wrong and I landed.

Trager and the young kid got scared. The mechanic thought my plane was covered with ice and so he wanted me to land. He was just scared. I bawled out Trager. I said, 'I'm going to go home. I'm going to leave you here with this airplane.' He said, 'No, you can't do it.' I said, 'The only way I'm going to go on to Huntingburg is if you get rid of that kid. I have no time to spare. I'm not going to screw around down here.' I made him send the kid home on the train. We were stuck there for seven days. We had a farmer plow the field down with a sledge because it was full of snow. Two big trees were at the end of the field and we'd have to clear them to get out. We tossed up to see who would take off first. I won the toss.

I had half the baggage and he had half the baggage. I took off and aimed right at the trees. I got in the air, rolled sideways, went around the trees, and came out. As I looked back Trager started flying towards the trees and did exactly what I did. We followed each other out and got into Albany. We stopped for gas in Albany. That night we got as far as Buffalo. We stayed overnight in Buffalo, and the next day we were headed for Cleveland.

It took us all day with headwinds to get to Cleveland. We were without lights over the airport. I was flying with my little tiny radio on my ear. I listened to the weather station. I was flying the on course and then I saw a green light flashing. The light signaled that it was clear to land. Trager was hanging on to me, following me closely in the dark. We had no batteries, or running lights, but Trager could see me. We landed in the field and caught hell for landing there. Cleveland Airport was a monster of a field. It was all grass except for the tower, some ramp space and hangers.

We took off the next morning and found the airport in Huntingburg, Indiana. It was windswept—there was nothing there but a tin hanger. We saw no sign of life. It was cold and raw and blowing snow. I circled around with Trager following behind me. All of a sudden, about fifteen cars started driving out from the local town. They knew it was us coming in so they all came down. We landed and everybody shook our hands. I was the chief pilot, he was the fixed base operator. It was around Christmas time. That night they took us to the local church where a woman sang *Avé Maria.*

It was a Germanic town. They had the Bunds in town. Bunds were groups of German Americans against the U.S. going into the war. They were called brown shirts in those days. They didn't like Jews. I said to Trager, 'They don't know you're Jewish. If these guys find out you're Jewish they'll never do anything for you.' He said, 'Oh, no, you're wrong.'

Trager's real name was Finklestein. They used to call him Finkie, until his family changed the name to Trager.

They paid him money but they didn't know that he was Jewish. I said, 'You're going to get into trouble here. They'll get you to open this airport, and set up an operation. Then Schwartz' son, the pilot, will come back from the war and they'll take this from you.' I said, 'I can't see anything but trouble here, I'm leaving for home. You can get double your money on the airplanes.'

I took the train and went back home. A week or two later, Trager took my advice and sold the airplanes. He had paid $1600 for each plane. He went into the next city, and he sold both airplanes for $3500 each. He got $7000 for a $3200 investment.

I WAS HOME A WEEK when I got a telegram. It said: *If you wish to be in the Air Force you are to go to Manchester, New Hampshire, to Grenier Air Force Base, and take a new medical, and another flight test.* Every 90 days the medical expired. It had been that long since I went down to Baltimore.

I was home in bed with a cold on Essex Street in Malden. Trager and his friend Cliff Robbins asked if they could go with me to New Hampshire to take the tests. I said okay. The next day I got out of bed and we drove to New Hampshire. I presented my letter and they gave us medicals. Trager didn't pass. He was colorblind. He tried bribing the sergeant, and an officer in the next booth was listening. The sergeant accepted the bribe. Then the M. P.'s went in, grabbed Trager, and threw him off the base. Cliff Robbins and I both passed and had no problems. Robbins became a captain in the Air Force later on.

JULIE IN HIS 2ND LIEUTENANT'S UNIFORM AT NEW CASTLE ARMY AIR BASE, WILMINGTON, DELAWARE.

For the flight test, an airplane had to be flown in from Mitchell Field in Long Island, New York. They called me two days later and I went back to Manchester. A Captain Mitchell had come up from Mitchell Field on a BT-14. The captain said, 'You fly it from the back.' He didn't trust me to fly it from the front. We took off and climbed out to altitude. He asked me to do a few aerobatics. I did the aerobatic routine that I had taught to all my students. He said I could go in and land. Once we landed he said, 'Let's switch seats now.' I switched seats and I did another full aerobatic routine. I landed again. Captain Mitchell said, 'You'll hear from the Air Force.'

Two weeks later, I was in Boston at the State Guard terminal. Florence called me that evening and read me the telegram. It read: *You are now commissioned to 2nd Lieutenant in the Army of the United States as pilot. Get yourself sworn in by your Justice of the Peace and report in 24 hours to New Castle Army Air Base in Wilmington, Delaware.*

Twenty-four hours would bring me there at night. All officers had to buy their own uniforms. The next morning I went into Rosenfield Uniform in Boston and I told them about the telegram. I got a full set of uniforms including the insignia for the outfit. That night Uncle Iz and Florence drove me to South Station in a snow storm. The trains were so busy, you couldn't get a sleeper for $1000. When I showed them my orders they gave me a sleeper. I got on the train at midnight. Florence and Uncle Iz had an awful time getting home in the storm. They had to leave the car on Belmont Hill. They trudged through the snow to get back, it was such a storm.

It took over seven hours on the Washingtonian train to get to Wilmington, Delaware. A car picked me up at the station and took me to the base. I had to sign papers all day long for family health insurance, life insurance, and the whole works. I filled out papers until about 4 p.m..

At the end of the day I thought about how I'd left everything in Boston. I had to wrap up my part of the trucking company and make sure everything was all set. I went to see the adjutant, the chief of office at the base, and asked if I could get a leave. He said, 'Sure, how long have you been here?' I said, 'I came in this morning,' and he went crazy. He said, 'You can't have a leave, you're here three months before you get two and a half days off per month.' I said, 'I've got to have at least a couple of weeks.' He said, 'I'll give you four days.' I accepted the four days and got a ride back to the train station. I caught the train as it was pulling out of the station. I hung onto the rail with a suitcase in my hand headed for Boston.

I got into Boston about midnight. I got a cab to Malden. We had closed up our apartment the day before. Florence had moved in with her mother and father with the twins on Essex Street. Three of her brothers were already in the Army so they weren't home. Her mother had eight children but they were all scattered around. Some were in the military, some were living elsewhere. I rang the bell. It was late at night. Florence got madder than hell at me. She thought she wasn't going to see me for months, that I had gone off to war, and there I was.

My brothers-in-law decided to close Beaver Trucking down. They sold the trucks three months after I left.

They didn't want to run it until the war was over. Izzy and Paul went to work for the Charlestown Navy Yard.

At the end of the four days I went back to Wilmington. I was in the alert room for our squadron. I was assigned to the 12th Ferrying Squadron and I wanted to fly. I went over and saw Captain Mitchell, who was the Operations Officer. I said, 'I'd like to fly an airplane.' He said, 'Hey, there are people in here who haven't flown for three months. Who do you think you are to fly when you first get here?' I was sitting there wondering what I was going to do if I wasn't going to fly. In walked Jim Marley, one of my former students who was a pilot. He had worked in the Boston Fire Department and was now a 1st Lieutenant. I talked with him for a while. Then in walked George LaCroix, a pilot from the Framingham Airport. I had flown with him to Maryland before the war. He was a Captain. George said, 'Would you like to fly? I'm headed to transition school and I want to fly a BT-13.' I said I would like to go with him. I flew for about an hour and a half. I only needed to fly three hours a month to get flight pay.

A 2nd Lieutenant was the lowest officer grade. I got $167 a month and 50 percent of that was flight pay. I got subsistence pay, $42 a month, and a room allowance, $32 a month. I assigned everything to my wife and kids and left myself $20 a month to live on. If I flew somewhere and stayed overnight, I was paid $6 extra a night. I flew at least ten overnight flights a month. I made $60 to spend on food plus the $20 which paid for my room.

The next day George LaCroix called. He was flying a UC-78, which was similar to a Bamboo Bomber. He asked me if I'd like to go. I had never flown a twin-engine plane, so I said yes. I went off with him for two hours, and I came back with a Form 1 for flight pay. I only had 3½ hours and I already had my flight pay for the month. Captain Mitchell got madder than hell. He said, 'You're a goddamn brownnose.' I said, 'What are you talking about? He wanted me to fly with him and I want to fly. What's wrong with that?' He said, 'The guys in here

OFFICER'S PAY DATA CARD 1st Lt.,
JULIUS (NMI) GOLDMAN, O-511760, AC

Over 0 years' service 2nd pay period years completed
, 19

Monthly base pay and longevity $ 166.67
Additional pay for Flying 83.33
Rental allowances 75.00
Subsistence (30 day month) 42.00
Date 6/20/44 Total, $ 367.00
Dependents (state names and addresses):
(wife) Florence Goldman,
165 Essex St., Malden, Mass.
Evidence of dependency (mother) filed with voucher No.
, 19
Accounts of
Allotments, class E, $ 160.00 $ $
Insurance, class D, $ Class N, $ 7.10
Pay reservations, class A, $
Other deductions, $ Class B $18.75
Subsequent changes in above data with dates thereof:

Changes affecting pay will be entered here and maintained up to date.
W.D., A.G.O. Form No. 77
March 20, 1942

THE ADDITIONAL PAY FOR FLYING COVERED JULIE'S ROOM AND BOARD OFF BASE. THE REST OF HIS MONTHLY CHECK WENT TO FLORENCE AND THE TWINS IN MASSACHUSETTS.

haven't flown for three months.' I didn't think much of it. I said, 'That's just too bad, I'm going to keep on flying.'

George called me and asked if I wanted to be an instructor in his transition school. I said, 'Sure, as long as I can fly. That's what I want.' I became an instructor taking out young lieutenants who came out of Air Force schools. I did that for a while and then stopped. I was more interested in flying across the country. Meanwhile, I started to transition myself on bigger airplanes. I got checked out in an AT-6, an AT-9 which is a twin-engine airplane, and a Loadstar, or C-60. Then I wanted to fly the Thunderbolt. I was told I could do that. I got checked out in a Thunderbolt but I couldn't get my Pursuit rating. I'd have to go to Instrument School first. So I went. An instrument rating is identified with a P. Flying light airplanes requires a 1P rating. Flying a light twin requires a 2P rating and a 3P rating is needed to fly pursuit.

Instrument School was right in Wilmington. I put in 50 hours in a BT-13 link trainer and 50 hours under the hood. The hood is a canvas shield that goes over the windshield so you can't see out. I could only see the instruments in front of me. They still use that in training now. Sometimes I'd have to put a set of glasses on my head, so I couldn't see out. The glasses had a shield, which was like a little hood over your eyes. If I picked my head up and looked out they could see that I was cheating. When I got in the airplane the canvas covered all the windows, and I had to fly just by instruments.

I was sitting in the alert room not doing much and I heard them announce ten names to come and get on an airplane to the Fairchild Factory in Hagerstown, Maryland. They were to go and pick up ten PT-19s and take them to Huskogie, Oklahoma. They kept calling the name of one guy and he didn't report. I went to Captain Mitchell. I said, 'Look, if you don't find him, can I fly on this flight?' I'd never been on a flight before across country. He asked if I was checked out in a PT-19. PT-19s were open ships. I hadn't been checked out in one but I'd flown it while teaching aerobatics. He said it didn't matter, I needed the transition check out. I said, 'If I get checked out before the airplane leaves for the factory, can I go?' He said, 'Yeah, I'll put you on it, but if you don't show up...'

I ran and went into transition on the flight line. I said, 'Do you have a PT-19 that I can get a quick ride in?' I said, 'I need a check to prove that I can fly this.' He said he had no PT-19s there. Just then a PT-19 came in on a cross-country trip. The pilot got out and went to lunch—it was just before noon. I said, 'Well. what's that?' He said, 'It doesn't belong to us, it belongs to another outfit.' He said, 'Hey, he isn't here anyhow. Let's take it.' A PT-19 was an open ship, with no hatch. The pilot was protected under this little windshield but the wind was blowing cold. It was the wintertime. It had two seats—one in front and one in back. They assigned an instructor to me and I went around the field with him once. The instructor said, 'I want you to fly around twice more so I can say that you passed the transition.' He got out of the airplane. I made five takeoffs and landings. I came back and they gave me hell. The pilot was back from lunch and he wanted his airplane. I ran with my parachute back to the headquarters.

When I brought the chit back to Mitchell, he said, 'How'd you do that?' I said, 'Well, I got it.' The chit said I was qualified to fly the plane and they assigned me. I was number ten man on a flight of ten. We flew to the factory in a Boeing 237D. Off we went to Maryland, and we all were assigned new airplanes. We left that afternoon, for Lynchburg, West Virginia. I just followed these guys, watching them. There were two flight leaders, Lobinstein, a 2nd Lieutenant, and Flight Officer Smith. I was Tail-end Charlie on Smith's flight. I followed them to Lynchburg, staying in formation, and landed there at a fixed-base airport.

It was late in the winter and it was very cold. We had on heavy flying suits. We got up in the morning early, had breakfast, and got out there. Nobody could start their airplanes. I had learned how to start airplanes in cold weather from living in the north. I had trouble starting my old Curtiss Sedan from time to time so I went over to the local mechanic and got a spark plug wrench. The PT-19s had an in-line engine, with six cylinders hanging down, and spark plugs on each side. I took a panel off and took out all six spark plugs on the left side. I put them on a radiator in the airport office. I went outside, checked out the airplane and when I came back the spark plugs were nice and hot. I put them all back in the engine. With one turn of the crank it started. Now everyone wanted me to start their airplane, I was the only one that got my plane running. I told everyone what to do and we got them started.

It took us a better part of the day to get off the ground. That night, we only got to Charlotte, North Carolina. The next day we were on our way to Spartanburg, South Carolina and a train was going by below. One after another the other pilots buzzed the train, like they were dive bombers—they were frustrated fighter pilots. Of twelve guys, one was sitting up there flying. That was me. The others were all buzzing a cow, a chicken, anything that moved, they buzzed it. They didn't know where the hell they were. I was sitting there at 3000 feet just watching them, my head out just flying. Every so often they'd fly formation with me. Somebody would try to tap a wing just fooling around. If one wing hit the other wing too hard, both planes would go down. They were stupid, playing. They thought they were hot shot pilots but I wasn't interested. We went from one station to another that way, all the way down into Louisiana and back up into the airport in Muskogie, Oklahoma.

Most of the guys didn't say anything to me, even the flight leaders, Smith and Laubinstein. We delivered the PT-19s and then we went to the nearest airport. We bumped ten people off of an American Airlines flight. We had the highest priority other than the general staff in the Air Force. I froze the side of my face on that trip to Oklahoma. I went to the hospital and the doctor said it was going to be scarred. He also said, 'Furthermore you have tonsillitis.' I said, 'I had my tonsils out when I was a youngster.' He said, 'I don't care what they did with you then, you have tonsils and we need to take them out.' I said no. The captain there, the medical officer,

said, 'I'll give you two weeks leave.' I said, 'Sold, you can take them out.'

That's when I met Jack Brissey and Isabelle Bornstein. Isabelle Bornstein's husband had worked for us in the trucking business in Lewiston, Maine. When the war started, Isabelle joined the Air Corps. She was a nurse in the hospital. Jack Brissey was getting his tonsils out, too. She and Jack met there and fell in love. She divorced her husband and Brissey married her. Her nickname was Shorty. I stayed in the hospital for four

horrible days, and I had a hemorrhage. I went into the headquarters to get my two week pass. The medical officer was transferred to England and nobody knew about me getting two weeks off. I raised so much hell, and again they gave me four days off. I came home to visit and then went back to my base.

THEY NEEDED PILOTS to be moving airplanes all the time. I wanted to fly as much as possible, for the money. When I got back I thought I better get a trip or two, and I went right to Operations in Headquarters to see Major Matz. I didn't talk to Mitchell because he didn't like me. Matz said, 'You got a lot of flight time right out of that trip.' I said, 'I want more trips.' He said, 'You had plenty of flight time, you were the flight leader.' I said, 'No I wasn't, I was Tail-end Charlie. I took off and went to the different stops and they followed me.' He said, 'If you want to fly, flight leaders fly all the time. Everybody else gets what's left but flight leaders are in demand.' I said yes. From then on I had no trouble getting flight time.

The Air Force had me flying B-26s, the bomber, out of the Martin Factory in Baltimore. The B-26 Martin Marauder was called the Flying Prostitute. It had very small wings, and was flying with no visible means of support. When they changed the wing it worked better. They built the airplanes and my outfit picked the planes up at various factories all over the United States. We'd fly them back to the East Coast to our base where they could be picked up easily and used in the war effort.

I was a 2nd lieutenant leading guys who were 1st lieutenants and Captains. To be a flight leader I had to know where the hell I was going, especially without a radio. Sometimes I had just a map and an open ship. I had already checked out with the Thunderbolts and AT-6s and AT-9s. I got my instrument rating and I was now a 3P pilot. I wanted to go through Pursuit School. I saw an opening and asked if I could go. I was told I was too big of a guy and should be flying big planes. I insisted. They said okay and gave me an appointment to go to Pursuit

School. They put me on a train in Philadelphia. I rode the train for three days and nights to Palm Springs, California.

I had a sleeper. I was the most popular guy on the train. Lots of girls wanted to use my sleeper. They'd promise me anything, but I wouldn't let them use it. I was using it most of the time. All seats were jammed full. There were no card rooms or bars on the trains. Every train was jammed full of people from one end to the other. Palm Springs was just a desert, it hardly ever rained there. It rained one time and my bunk house got flooded. There was nowhere for the water to go. I put my hand on the floor from my cot and it was in the water.

I went into Pursuit School where I flew P-40s, P-39s, P-51s, P-47s and various models of aircraft. I came out of school as a flight leader. I had to learn instruments for all the different planes I flew. I spent 100 hours under the hood in a link trainer. My results were traced on a paper. The teacher told me where to go and I had to fly it. I was hooded in, and I had to fly the demo. If I didn't do it right I'd start spinning around in circles, meaning I was dead. Then I put in 50 hours in a real airplane sitting in the back seat with the hood up. I had to make approaches and land under the hood.

When I finished school, I had a flight of P-51s, or Mustangs, to lead to the east. The P-51s usually had no radios to speak of. But these airplanes had the first of the VHF radios, with four channels, called a BC-522. It was a very heavy radio. It probably weighed as much as a person. It was sitting in the back behind the cockpit. The pilots could talk to each other, over the four channels—

A, B, C, and D. One airplane to another could talk. Some towers had VHF as well. Before that, all you could do was listen to the communication and navigational equipment. Low frequency radios were used but they were very cumbersome. VHF was brand new.

Six of us took off from Palm Springs and landed in Tucson, Arizona, for fuel. There was no fuel but they said we could get fuel at a training base, about 75 miles south of Tucson. A special octane fuel was used for P-51s. We all got back into our airplanes, got down there and sure enough they had fuel. The base commander came over to me and said, 'These kids here haven't seen a real airplane since you guys came. Give us a buzz job.' That's all we needed. We practically tore the place apart.

We arrived in El Paso that night. Night flying was limited. I had night flying on other planes but not in a P-51. Fort Blitz in El Paso had a long runway that was closed. The civilian field was open—it had a 4000 foot runway. It was at the side of a mountain. I was told to look for a well-lit statue on a cross on the mountain. I saw the cross and knew it was the right field. When the power is cut on a P-51 at night, the long engine flame turns yellow. At full bore it's blue but with yellow I couldn't see anything. I was worried about my kids getting into trouble. I talked to the kids and one after another I knew that everyone was okay.

We landed, taxied in and parked. We had the planes refueled and stayed overnight. I was standing at the dispatch desk when somebody banged me on the back. I turned around and it was Captain Pompernick. He

had gone across to war with his outfit. They were taken out of Logan, and my brother-in-law Vic was with them. They went through Casablanca for the invasion. They went through part of Africa and then on to the invasion of Italy. They went into Anzio, and took a terrible beating there, where Vic was wounded. Vic never had the wound taken care of, he took care of it himself. He didn't want the Purple Heart because his buddies were killed.

I said, 'Well, where is Vic now?' He said, 'He's back there with his buddies. I came back here to Fort Bliss to form a new outfit. I wanted to bring him, but he's just happy with his buddies there, what's left.' Pompernick had just arrived when he saw me. I said, 'Why didn't you try to bring him home?' He said, 'I couldn't. He was just interested in staying with his friends.' He added, 'He could of been a captain by now.'

I stayed overnight and in the morning I got all my guys ready. I was talking to them on the VHF radio. All the planes were lined up, and each guy took off. I had been taxiing so long that I fouled up the spark plugs and had trouble with the tail wheel. The tail wheel in a P51 unlocks by the stick being pushed forward to the firewall. I pushed the stick forward and it wouldn't unlock. A cable was improperly adjusted. I had to open the throttle and bounce the thing around, until I got lined up on the runway. The kids had all gone, headed for Abilene.

I got in the air about fifty feet and the engine quit. I was headed in just the right direction. There was a runway at right angles to me. I pulled the nose up. The gear had just started up, so I started the gear down. There was still pressure from the prop filling. I rolled it onto the runway and I landed. I was lucky I didn't kill myself.

The engine was still flipping over slowly. I called the tower on VHF and I told them what happened. They said to come on in and they'd check it. I kept running the engine at full bore as I was taxiing. I said, 'I want to try this again but I'm going to pull a lot of power before I let the brakes go.' I got up to 38 pounds of manifold pressure and I let the brakes go. When I got a third of the way down the runway, I had it at 55 pounds of manifold pressure, full power. It held so I kept on going. I got in the air and upped the gear. As I did that I heard a bump and she started to skid sideways. The adjustment on the landing gear doors was out of whack. The wheel closed outside the doors, so I was skidding. I slowed it up to 180 mph and dropped the gear. This time the wheels closed up and the doors closed properly. I was on my way.

I had a bad head cold and I was flying at about 18,000 feet. I didn't have any oxygen. When I got close to Abilene I started down in a hurry and my ears blocked up. When I landed I was in agony. I taxied out and they parked me. As I climbed out I put my foot into a step. I slipped and fell onto the wing on my back, and then onto the ground. The whole side of my airplane was covered with oil. A vacuum tube on each side of the engine was improperly adjusted. If one tube is shorter than the other, the oil is sucked out.

I didn't know what happened. I was lying on my back stunned. I saw fluid coming out of the big scoop in the belly with three radiators in it. One has glycol for the engine, one has glycol for the supercharger, and one is for the oil. Everything was leaking, and my head was one tremendous headache from my ears.

I picked myself up. I told them I had trouble with the tail wheel and they fixed the airplane all up. I told them about my ears. They put me in the pressure tank and took me up to 18,000 feet and then lowered me down slowly. It took an hour and a half to get me down to sea level and then my ears were open.

At night, oxygen was to be used from the ground up. In the daytime, oxygen was used from 10,000 feet up. I was young and stupid. I flew up to 18,000 feet without oxygen. I didn't bother with it until we had oxygen training. Ten guys were put in an oxygen tank. Each of us had a mask on. The oxygen was removed from the air as if we were climbing in altitude. We got up to 35,000 feet with oxygen masks on. Then we had to take the mask off and write our name and serial number as many times as we could. I pulled off the mask and started writing. I thought I wrote it five or six times. I wrote it once. The second time I wrote my name with no serial number. By then I hit the floor and they put the mask back on me.

I called Trager from Abilene. He was in Sweetwater, Texas, teaching girls to fly for the Air Force, as a civilian. I just called to say hello. He said, 'Don't go, I want you to come here and stay overnight.' I said, 'I can't, I'm on my way east. I have a terrible cold and I can't come there.' He said, 'Don't go, and I'll come there.' Sure enough, in about two hours he arrived in Abilene in an old

Studebaker. We spent the night talking and he did everything he could to convince me to stay.

The next day I took off. Now I was chasing my guys who were in Tulsa, Oklahoma. A guy named Taggard was my subflight leader. I went racing to Oklahoma and found my flight in Tulsa. I landed there, got everything squared away and we decided we'd take off in the morning.

That morning, a C-47 took off ahead of us, covered with frost. He got off the ground and wasn't 50 feet in the air when he stalled. He went down on the ground, rolled through a fence, and hit a house, moving it off its foundation. He banged up the nose of the airplane and the engines. He was lucky he didn't get killed or kill anybody in the house. Frost destroys wing lift. It changes the contour of a plane's wing. I was the first one to take off. I was in the air and I pulled on the handles to lift the landing gear. The gear didn't move.

I got a call from someone in my flight. 'Hey Goldman, you forgot the landing gear. You know that's a no-no.' I said, 'I didn't forget anything. You guys take off. Taggard, lead them on.' I went back in and landed. It took me two days to find out what was wrong with that airplane. No one there knew a thing about P-51s. I had gone through flight school, I learned the system and I had a general idea. I got a hold of a cable that turns a three-way valve and when I pulled on it the gears went up and the doors went down and there was a crash. I almost got decapitated in there. One cable was loose so they adjusted it.

I took off and headed to St. Louis, which was on top of an overcast. I got to about 5000 feet up and was listening on the radio for navigation. I wanted to make an approach on a four-course range. Everything was clear. I came down out of the clouds and there was the airport, Scott Field in St. Louis. I landed in there, taxied, and there was a P-51 sitting there. It was Taggard. He was having trouble with his gear and the kids had gone on without him.

I taxied in and asked for fuel. Taggard wanted to fly with me once he got his gear fixed. We made an instrument takeoff in formation. I could see him behind me and then I couldn't see him anymore.

I figured he was flying slowly and he'd come out. On the VHF radio, he said, 'I heard a thump.' He could get the wheels down to land the plane but he couldn't get them closed. I said, 'You must have the same problem I had. The gear is out of synch with the doors.' He came out of the clouds and I could see one wheel and the door stuck together in a V. I said, 'You'll have no trouble, slow up and drop the gear and cycle it. I'll fly along side of you.' It didn't work. He was out of synch entirely. I flew circles around him all the way to Indianapolis he was so slow. I called the Indianapolis airport and told them that we needed help with the landing gear. They had somebody who could fix it and we went in and landed.

Once they fixed the problem, we were rather late. We went on to Cleveland, a huge grass field with a ramp. The wind was blowing so hard out of the west that we had to land perpendicular to the normal route, across the ramp. We were going about 110 mph but because of the

JULIE WITH NAT AND SHIRLEY TRAGER YEARS AFTER THE WAR.

wind the planes were only moving 60 mph on the approach. We were in formation, and landed on the ramp facing the tower. It was hard to turn the P-51s around with the wind but we did it. American Airlines said we could put our planes in their hangar for the night. We came out in the morning and the wind was still blowing like hell and freezing cold. They pulled the airplanes out using tugs, tails into the wind.

We started both airplanes at the same time. All of a sudden I saw my Glycol temperature going into the red. I realized that the wind was counteracting my prop fires through the radiators. I spun the airplane around and I tried to catch Taggard's attention. He had his head inside the cockpit. I kept calling him on the radio and I couldn't get him. Finally he looked up and saw me. He put his radio on. I said, 'You better spin around into the wind. You got the temperature on your glycol?' He said, 'It's in the red and it's steaming.' He spun around. He said, 'I'm afraid to go, I better have it checked.' He was going to Niagara Falls and I was going to Rome, New York.

They cleared me and I took off. I followed the flight plan for an hour and a half. I was flying at 6000 feet when I went into a snow storm. I couldn't see anything so I was on instruments. I was trying to talk to the Buffalo tower. I was coming up on the tower going like hell. They asked, 'Where exactly are you? We have a DC3 climbing through your area.' I said, 'I can't tell you, I didn't expect a snow storm here.' I started a 360 in the snow and I laid it over pretty good. I almost stalled the plane in the tight turn. I said to Buffalo, 'I'm going to fly

south away from the lake and then I'm going to fly east five minutes then north five minutes and I should pull around your airplane.' I flew the four-course range. There was no radar at the time. Scientists were just trying to develop it. You had to have educated ears. We had low frequency range to tell us which direction we were flying. At about four minutes I ran across the south leg of the Buffalo range. The tower said, 'He's flying on the west leg and you're clear of him.'

I turned onto the east leg and as I turned I came out in the clear. The sun was shining and it was just like a summer day except it was cold as hell. I intersected the

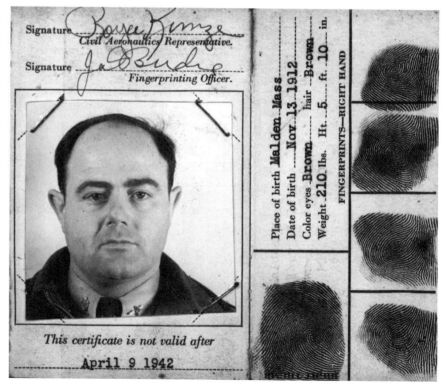

east leg and I flew down the east leg to Rome. I got to Rome and looked down. The field was all white. I could see the layout of the runways and I flared out the top of the snowdrifts which were about fifteen feet high. I was afraid I would roll the airplane over because I was going so slow. I gave it some power and it squatted down like a feather in the middle of the drifts.

There was an inch or two of snow on the runway and big drifts on the side. I came to an intersection and taxied in by the administration building. A sergeant came running out. He climbed up the wing and shook my hand. He said, 'You're the first pilot that's ever followed a flight plan into here that's here on the minute.' He didn't know I was all over the state trying to stay away from a DC-3. I hit my flight plan right on the button and that's where I left my airplane. I took the train to Boston never telling the base where I was. I was supposed to follow the rules but I snuck home to Boston instead. I stayed home with Florence and the kids for a day, or two. Then I went back to Delaware.

They never noticed. Every stop I made I had to send a telegram. However many airplanes I was in charge of, I'd send a telegram letting the base know where we were, and that we were remaining overnight. When I'd take off the next day, I'd send another telegram reporting back to base until I arrived home and the mission was complete.

I TOOK THE FAMILY CAR with me to Delaware. It was a Chrysler Windsor. I left it with Florence at first, and then I drove it down to Delaware to get around the base. I had bought it brand new from the factory at $900 and it was a wonderful car.

Florence came down with the twins to the base. The train station was packed with people. There just wasn't enough transportation with everybody traveling. Florence and the kids were up all night. When I picked them up at the train at 7 a.m. they were shaky from the ride. In the Air Corps, pilots hardly ever saw their families. I never saw them except when I'd come home from a trip. I'd sneak home as often as possible. If I went through Boston, I'd stop off. If I went through New York, I'd bum my way to Boston. I'd come into Manchester, and they'd pick me up. I had the twins' shoes hanging in the windshield of my car to let people know I was married. They were tiny white moccasins. I kept a curl from Marlene and Myrna in my wallet so they were never far from my mind.

One day, Operations sent me to pick up an L-4 Cub from the manufacturer. I flew to Lockhaven, Pennsylvania, and then had to fly the plane to Mobile, Alabama. I had just checked out in a P-47 and was used to a rather heavy airplane with lots of speed. As I headed down the runway in the L-4, it seemed like it wouldn't fly. But when I pulled back on the stick the plane shot up in the air. I headed for Middletown, Pennsylvania.

Over the mountains I lost oil pressure but decided to keep flying in the hope that the engine wouldn't quit. There were 40 mph headwinds. I struggled all the way down to Middletown. I landed there and the maintenance shop fixed the oil pressure loss. They found no damage.

I continued on to Charlotte, North Carolina, where I spent the night. I flew 12 hours against 35-40 mph headwinds all the way. By the next night, I had reached Dothan, Alabama, where there was an advanced AT-6 school. Again, I had flown 12 hours and the headwinds hadn't let up. I landed at Dothan just as they were having the graduation of the Senior Cadets. The Base Commander asked me to attend the ceremony as I was a military pilot. I didn't want to go to the graduation. I had nothing nice to wear with me, but I had no choice. I stayed up late, and then got up early to fly the next day.

I couldn't fly straight to Mobile, Alabama, since there were no refueling stops. I refueled at Maxwell Air Force Base. I refueled again at Eglin Field in Florida before flying west to Mobile. The final leg of the trip took six hours the third day. I had 30 hours total flying time in three days. Pilots would normally fly a maximum of five hours in one day. There were 35-40 mph headwinds for the entire trip and since my cruising speed was about 75 mph, I only made a ground speed of 40 mph. It was heartbreaking to see cars occasionally pass me on the highway below.

After delivering the plane, I took a commercial flight from Mobile to Washington, DC, where I caught a train to Wilmington. No seats were available on the train so I had to stand all the way to Wilmington. I had been gone from the base for three days. On the morning of the fourth day I walked into Operations, unshaven and looking a bit crummy. The Operations people figured they had gotten rid of me for at least two weeks. I showed them the receipt for delivery of the plane to Mobile, Alabama. I then asked for another trip.

FOUR DAYS LATER I was assigned to take an AT-16 from Wilmington to Northrop Aviation Plant in Hawthorne, California. An aviation mechanic from Wilmington was going home on emergency leave. I was to take him and drop him off in Joplin, Missouri. There were no radios in the AT-16. It had two full panels of instruments with an old straightaway compass on the floor of the cockpit. The airplane was going to Australia so I guess they could care less.

I left Wilmington early in the morning to fly to Cove Valley in Connellsville, Pennsylvania, for fuel. Cove Valley had no fuel. The AT-16 had the same fuel tanks as an AT-6. They held 60 gallons in each wing plus an emergency 17 gallons in the left wing. When you were down to the emergency fuel a red light signaled on the panel. I decided I could make it to Pittsburgh with the fuel I had. The weather was extremely poor with rain and a ceiling of visibility 200 feet above the trees.

I had run out of fuel in both my main wing tanks and was down to the emergency 17 gallons. I arrived in the Pittsburgh area at the intersection of three rivers. Without a radio, I could not talk to or see the airport with all the fog, rain and clouds. I had to make a decision as to what I was going to do. Looking at my aerial chart, I spotted a circle. A civilian field was about 20 miles northeast of me. I set my directional gyro for a bearing that would take me to it. I yelled to the mechanic sitting in back. I said, 'If we can't land in ten minutes, I'll pull up and we'll have to bail out.'

In just short of ten minutes, I spotted the airport in front of me. It had two runways and a series of buildings. I didn't hesitate. I pulled up the nose into the clouds, dropped the gear and rolled back to the airport doing a 180 degree turn. After landing on one of the runways, I taxied up to a nearby building and an Army Colonel came running out. I opened the hatch and he shook my hand, saying I was the first plane to land on his field. He didn't tell me where I was and I didn't ask. I didn't want him to know that I was lost.

I went to the weather station and talked to the 2nd Lieutenant in charge. He showed me on the map what the weather was like in St. Louis. I found out that I was just where I was supposed to be—at the civilian field at Connelsville, Pennsylvania. The field had just been built into a Sub Depot airport. The Colonel invited us to lunch and then we filled up with fuel. The young 2nd Lieutenant in Operations pointed out a cut in the trees to the west of the airport. It was a pipeline that ran all the way to St. Louis.

After thanking the 2nd Lieutenant, I took off to the west and followed the pipeline. As I flew, the ceiling of visibility gradually rose from 1,000 to 9,000 feet. I climbed in elevation with it. About half way to St. Louis, I thought I was low on fuel due to the head winds. I headed slightly north to Terra Haute, Indiana, to refuel. Again, they had no fuel. I took off for St. Louis in the hope that I could make it on my remaining gas. Again, I used up my main tanks and was running on the 17 emergency gallons. I stayed at a high altitude so that if my engine quit I'd have space to spiral down and land.

St. Louis was about 40 miles away. Being so low on fuel and with no radio, I landed at Scott Field without getting clearance. I spiraled down on the field and no one questioned me as I taxied in on the fumes. When I asked for 91 octane fuel, they said they had none. I explained that all I needed was 25 to 30 gallons to make it to the civilian field north of the city. They told me the General had 300 gallons for his UC-45 but I couldn't have any of it. I finally tracked down the General. I explained my situation and he said I could have as much of his fuel as I needed.

With 30 gallons of the General's fuel I was able to reach St. Louis Municipal Field to fill up. From there I took off for Joplin, Missouri. The weather was very good and I flew at 4000 feet with my hatch cracked about two inches. I held up my chart to show the mechanic where Joplin was and how far we had to go. The chart was sucked out of my hand, through the open hatch and was gone. I had no radio and no chart but I managed to reach Joplin and drop off my passenger.

From Joplin I had to get to Tulsa without a chart also. By guess and luck I reached Tinker Field in Oklahoma and started looking for a chart. I spotted a Civil Air Patrol pilot with a chart that would get me to Deming, New Mexico. I offered him $2 and he said, 'It's yours.' There was a Women's Army Corps captain at the operations desk. She asked me to take her to California because I had an empty seat. I knew it was against the rules to haul passengers with no orders so I refused to take her. She was unhappy and said it might be an interesting trip but I still refused.

I headed to Deming, New Mexico. I flew through the desert, past Palm Springs and went to San Bernardino. I had no radio so I circled until they gave me a green light to land. I rolled over and there was a four-engine airplane staring me in the face. I kept rolling and came in behind him. I was furious with the tower but I couldn't yell at them without a radio. I landed in San Bernardino for fuel. I flew down the valley from San Bernardino to Los Angeles where I was supposed to deliver the plane to the Northrop Factory.

I came over the last ridge and saw the biggest city I had ever seen in my life. I got over the mountains and there was Los Angeles. It went from mountain to mountain. Full of buildings, no open spaces. I couldn't find the factory and as I circled, two P-38s flew formation on me and tried to guide me south. I didn't want to go south. As I looked down, I saw a plane land in what looked like a big corn field. I cut the throttle and landed on the field behind him.

The runway was covered with pieces of rubber that looked like corn tufts, pasted on for camouflage. The government was worried about the Japanese coming in and bombing that area because of the recent bombing at Pearl Harbor. They camouflaged everything, the buildings and the runways. All the buildings had nets over them. It was Los Angeles International Airport.

I taxied up to a big building that looked like a barn and

a sergeant came out. I asked about the Northrup Factory. He said I'd just flown over it—it was on the other side of the field. Hawthorne was north of there. I was told to look for a couple rows of telephone poles. I took off and went around the field and I went down in between the rows of poles. The entire Northrop Factory was under nets. The strip between the telephone poles was the runway. Everybody was busy working. The nets looked like grass from the air. The Pacific ocean was at the end of the runway. The plane I landed was going to Australia and it was put on an aircraft carrier with a crane. Many planes were transported on cargo ships or aircraft carriers this way.

I was in my flight suit walking in and the women were looking out of the windows of the factory whistling and hooting and hollering. All of the guys had gone off to war and there were 2000 women in there. They shouted, "Come on up, come on up." The Northrop people took me to the Hollywood Roosevelt Hotel on Hollywood Boulevard. Across the street was the Walk of Fame and the Loehmann Chinese Theater. I wandered around a little bit. The next morning I flew back to my base. It was an interesting trip. It took about two weeks and more than 53 hours of flying time.

I WAS FLYING ALL OVER the country with my 3P pursuit rating. I was asked to go to Reno and fly C-46s. My ears perked up. I asked, 'What will that do for my rating?' It would give me a 4P. I'd have the second highest rating. I flew all the twins up to that point. A 4P rating is needed to fly a huge twin, like a C-46, DC-3 or C-47. There were various models depending on the size of the engines. I didn't know that they wanted to send me to the Hump, to fly the Himalayas. Nobody came back from the Hump.

I said I'd accept it and I flew out to Reno. That was a hard school. We were flying at night, flying in snow storms, so we'd be trained to withstand the Himalayas. I was flying various types of DC-3s—C-47s, C-49s, C-53s, and C-46s. A C-46 was the biggest twin in the world, it was called a Curtis Commando. The C stood for cargo. It had two 2000 hp Pratt & Whitney engines in it.

It was pretty tough flying. A few guys quit and we had numerous troubles with the airplanes. We were flying airplanes that didn't have good maintenance, and flying at night in storms. My training was marvelous. My instructor, Joe Walker, got me to do things that I never thought were possible.

One day, Walker, myself and my copilot, Schmidt, were out flying. Walker had me man an approach in a C-47 onto a strip in the mountains encased in fog. He said, 'You see that highline going down into the fog? Do you see it coming out near that mountain?' A highline is an electric line that goes through the mountains and follows the ground, like the wire that runs down the street between telephone poles. I saw one end of it going down into the fog, and then the other end of it coming out of the fog. Walker said, 'There's a strip in between there.' I said, 'So?' He said, 'You land in there.' I said, 'I don't think so.' He said, 'Are you afraid?' I said yes. He said, 'Well, get out of the seat and I'll show you.'

Schmidt was in the copilot seat. I stood up behind Walker while he made the approach. He lined up with the highline to his left, made the descent, went into the fog and just kept going down, down, down. He curved just slightly and we were on a runway. He said, 'That's how you do it.' I said, 'How did you know where the strip was?' He said, 'I knew where the strip was, now you know where it is. Now you do it.'

I got in the seat. I made the takeoff and I just missed the highline, I knew it was in the fog to the left and I came out. He said, 'Now go in and land.' I made my approach. Everything I did was due to watching that highline. Sure enough, first thing I knew I was on the runway. I rolled out and stopped, turned around and taxied back. I took off and I made three of those landings.

Walker said, 'Now, Schmidt, you do it.' Schmidt got in the pilot seat, Walker was in the copilot seat and I was standing behind Schmidt. I saw his entry. He was just drifting slightly toward the highline. I said, 'Schmidt, correct to the right.' I kept saying, 'To the right, to the right.' First thing I knew we were on the ground. One wheel was on the sand and one was on the runway. I said, 'We're running off.' Walker took the controls and kicked the airplane over to the right. He said, 'That'll be enough for today. Don't tell them at the base that we did that. That isn't part of the curriculum.'

We were making approaches on a four-course range in between the mountains. We had the hood up so I couldn't see out of the plane. We flew with instruments and followed the Morse code sounds of the range. We

made our approach on the north ridge of the standard four-course range. We went through the station and onto the south leg, did a procedure turn and then came around. We went back up north and made another procedure turn. We were heading down all the time in between the mountains. Our final leg was down to 7500 feet and we started at 11,000. We made the approach. I practiced this several times and I did it both in a C-46 and C-47.

One day we made a takeoff headed for Mount Donner Summit, a huge mountain west of Reno. It was late in the afternoon and we just broke ground. There

SAUL VISITED JULIE AND THE FAMILY DURING THE WAR.

was a crew chief, Joe Walker, and myself in the airplane. I made the takeoff. The airplane, a C-49, was heated by a water boiler behind the copilot. We had all the power on, and both engines were running full bore. Then Joe shut down the left engine to train me in a single-engine procedure. At the same time the boiler exploded. The boiler was the only way to heat the plane. A coil wrapped around the exhaust tank heated the water. The water pipes attach to the boiler and a radiator. There was a nose vent on the radiator which opened to let the boiler cool when it got too hot. The crew chief didn't open the nose vent. It got so hot it exploded and blew the little boiler apart. There was steam over everything.

I'm under the hood, and the instruments and windshield are covered with steam. The rear window of a C-47 slid back and forth. I slid the window open trying to see where the mountain was and I couldn't see very well. The hood was up on hinges and it was canvas. I tore the hinges and the canvas right off, but I still couldn't see through the steam on the windshield. I was trying to look out and see where the face of Donner Summit was. My one engine was going around and I could see out of the side window a little bit. I wanted to keep heading to the left because that's where I could see. I did the single-engine procedure. I cut the mixture and pulled the throttle back and put the prop into feather position.

I reached up to the two engine switches. I hit the wrong switch, turning off the good engine. There was silence momentarily. It seemed like an eternity. I turned the good engine back on, and it came in with a roar.

Walker looked at me and said, 'That's how guys like you get killed.' He was sitting there all the time watching us, letting me handle it, because that was what the training was all about. The fact that the boiler exploded was something that added to the problem. I quickly switched the engines on again and we kept on going. It was quite a situation for me but I kept it flying. I asked Joe if we should go back to the field. He said, 'No, maybe this is an actual flight. You have to continue.'

Walker kept on my back. He started off with five students. By this time there were only two left. Nobody would fly with him. 'The other three were no good,' he said. They didn't want to fly with him because he was so goddamn tough. Schmidt stayed and I stayed. We were the only two left out of the five. He was putting more and more pressure on me all the time. As soon as I thought I handled something well, he put more pressure on me. I'd never fly in the same kind of airplane. I would be in a C-49, C-53 or C-47. They were all the same basic model except they had different engines, and the throttles were different. You had to know the differences between the planes. Even the doors were different. The C-47 was a newer airplane too. The 49s and 53s were old airline ships. They were taken away from the airlines because there weren't enough airplanes for the military. The C-47 was an airplane powered by Pratt & Whitney.

I WAS FLYING C-46s and also going in the link trainer. I had trained in Delaware, Palm Springs, California and in Reno, Nevada, under the hood. I did the link trainer and then used the hood in the airplane. That training kept me alive. It was hard at first. It was very tense, but I got more relaxed eventually. I flew the biggest plane in the world landing it under a hood, making landings just watching instruments and touching down. Day after day I was getting an hour or two in the link. Link training had to be current for every pilot so the person was prepared to fly airplanes using instruments only. At the same time I was learning how to fly different airplanes on two engines.

One day Walker and I and a crew chief took off for Salt Lake. It was in the evening, just after sunset. We went to a depot in Salt Lake to pick up a load of parts for C-46s. We went to Alco, and then to Ogden, and from Ogden to Salt Lake, all in the mountains. We landed in Salt Lake and they loaded us up with heavy parts. They didn't put any fuel aboard. It was a nice night, but very cold. The airplanes were having problems with their tanks. We were told not to turn on the gasoline heaters in turbulence because the heaters were in the wings. There were three long gasoline tanks in each wing. When the wings would bend with turbulence, it would open up a seam in the tank and the fuel would go out. If the heaters were turned on, it could blow the wings off. It was a rough ride without any heat.

I had on a fur hat, nice boots and good gloves but I was wearing a light flying suit. It got cold. My elbows and my knees got stiff. Everything was stiff except my feet, my hands, and my head. In Salt Lake, a sailor said, 'Lieutenant, can you give me a ride to the west?' I

said, 'We're going as far as Reno. You can come with us.' He said okay and got in the airplane. I never even thought of telling him he'd have no heat before he got in. The four of us took off in the plane, loaded with parts. By the time we got to Alco it was snowing like hell at altitude and we were flying high. I got to Reno, made the double-shuttle approach, descending all the time in between the mountains in a blizzard. My first mistake was to put on the landing lights.

The snow reflected in my eye. I couldn't see a thing. I was down low over the civilian field. This strip was only 4000 feet long, and 4000 feet above sea level. There was a 600 foot knoll on the right side of the field. I had to make sure that I didn't move towards that knoll. Joe said, 'I think we're close to the hill, I think we're close to the hill.' I was trying to get onto the runway and I missed it. I talked to the tower and they said they saw my wheels

THE C-46 WAS THE LARGEST PLANE USED IN THE MILITARY FOR CARGO.

go by. I saw the engines had cooled down a little bit. I put the coal to it, and I opened up the throttle. The engines were shaking and twisting and throbbing and I had everything wide open. The power took hold and I kept it at full throttle until I reached 10,000 feet. I climbed up to 13,000 feet before I even heaved a sigh of relief.

The tower said, 'You're not getting in here tonight.' I said, 'Where do we go? I was told to go to Sacramento, over the mountains and into San Joaquin Valley. I didn't have any extra fuel. Joe said, 'This is the kind of situation you're going to run into in the Hump so you better learn.' I said, 'Well, we're going to stay up here. We'll have to jump if we run out of fuel.' We came over the Donner Summit in the snow. As we went over the top, it was just as if somebody turned on a light switch. It stopped snowing, and we looked down into the valley. It was five o'clock in the morning. You could see lights down in the valley. I cut the power down and sunk into the valley. I landed in Sacramento at the big base there. I taxied in and we got fuel. The sailor got out of the airplane and said, 'Do you mind if I kiss the ground?' I said, 'I'll kiss it with you.' We bent down and kissed the ground together. I was sure that we weren't going to make it that night. And Walker, nothing bothered him, nothing.

JOE WALKER HAD ALREADY been to war. He had been sent to Ku Ming, China. Walker worked in the air transport command of the China division. He and other pilots were on the Hornet, the aircraft carrier. Jimmy Doolittle and his men flew off the Hornet in B-25s for the Tokyo Raid. Walker waited on the carrier and none of the airplanes ever returned from the mission. They all ran into trouble. The group successfully bombed Tokyo and then ran into bad weather. They were low on fuel. All but one crew parachuted out of their airplanes. The one crew that stayed flying landed in Russia. The others worked their way through the jungles and got home. Walker was one of the best pilots flying C-46's in the Hump. He came back with a lot of medals. He was sent home to be an instructor at Reno. He was crazy as a coot but a very good pilot.

In Sacramento we went in to get something to eat. Afterwards we still had to get back to Reno. It was about 7 a.m., with bright, beautiful weather. The crew chief came charging up to us and said, 'Hey, we found a magneto out cold on the right engine.' The magneto fired the ignition system. There were two mags in each engine. I said, 'Well, we'll have to get it fixed.' Joe said, 'Make believe you're on the Hump.' I said, 'Why should I make believe?' He said, 'Because you're not going to get anybody to fix that mag. You gotta go.' If the other mag failed then the entire right engine would quit. With the load we had aboard, 5000 feet was the highest we could go on one engine. We had to go up to 13,000 feet to go over the Donner Summit. Joe said, 'Well, that's what's going to happen to you on the Hump.' I said okay and we took off. If we lost an engine we would have to turn and go down the slope or we could hit the mountain.

We came up over the summit and looked down at the Reno Army Air Base. It glistened, beautiful white, everything was white, sheen, smooth. It had sleeted on the runways and it was solid glass. I made the best landing I had ever made. I landed the plane just like a feather, taxied in, and parked. All the taxiways were on a slope. I shut the engines down. The ground crew didn't have a chance to get the spiked chocks behind the wheels. The minute I stopped and shut the engines down my plane slid backwards and nobody could stop it. It slid off the taxiway into the grass. Nothing was damaged but it was stuck in there. We had to get tugs to pull it out.

IN ORDER TO GRADUATE Reno as a 4P pilot I had to take a final check flight for so many thousand miles. Walker set up a cross country for me to fly from Reno to Sacramento Valley to San Bernardino to Los Angeles, and back to Reno. We took off at night and climbed up over the Donner Summit out of Reno. I took it to a high altitude. It was very cold and the heater wasn't working. There were the two of us and I had the hood up. Walker was tired and went in the back to sleep.

I was heading towards San Bernadino. I thought there was no sense in having the hood up because no one was being an observer. I got over Donner Summit, and called Air Traffic Control. I got permission to descend to 7000 feet. I pulled the hood down and looked out and see if I could see anything. It was pitch black out the window. Then I saw some lights right square in front of me. There was an airplane coming at me.

I rolled to the right and he saw me at the same time and he rolled to his right. We passed each other by about 50 feet. He was climbing to go up over the summit and I was descending. The tower didn't know where we were. They cleared me down and they cleared him up. I called air traffic control and told them what I thought of them. I was in a C-47, a twin-engine plane and he was also in a twin-engine. It was luck that I happened to look up and the other pilot happened to see me before we had a collision. That was close.

I went down to San Bernadino, and on to Los Angeles, then we refueled and went back to Reno. I got back to the base and they told me I was graduated as a 4P pilot. I said, 'How could that be?' I had 50 hours of link training but only 32 hours of flying. I was supposed to have 50 hours of flying. They said, 'You passed all the tests and you are doing so well that we want to graduate you.' I said, 'I don't want it. I want to get what everybody else has gotten.' I knew I was going to the Hump and I wanted as much experience as I could get. Joe said, 'I'll let you fly five hours in a C-46 after you take the check ride and graduate.' I agreed.

I took a check ride with a Captain Gothrup, and it was the roughest day we had flown in the months I was there. The airplanes were restricted from flying because it was so rough in the mountains. The wind was bending the wings. If you bent the wings too much the gas tanks would split. Gothrup said let's go, so I went out with him and a crew chief. By the time we climbed up to altitude, Gothrup was taking oxygen from a tube—he didn't have a mask on. I was too busy to take oxygen and the crew chief was sick. It was one tough ride. I made all the approaches and flew for more than four hours. I came in and landed and they gave me marks, and passed me.

For the last five hours of training Walker promised, my copilot was Schmidt. The two of us went out one night in the snowstorm making approaches in the C-46, doing just takeoffs and landings. When we had five hours of training, we came in and that ended it.

When the war ended, Walker went to Edwards Air Force Base as a test pilot. He was out flying an F-100, or 102, which is a chase plane on the B-70 Valkyrie. The Valkyrie was a huge airplane that sat on high nose tricycle gear. They were taking moving pictures of the flame off the engines in the Valkyrie, flying in close formation. He sucked it in real close and got caught in the vortices. Walker rolled into the Valkyrie's fuselage and the whole works crashed. Everyone was killed.

Joe Walker had experience but when you are flying at high speeds the wind peels off the end of the wings and it creates vortices. He flew in underneath this Valkyrie with this fighter which didn't have such a big wingspan and as he flew into it, the vortex off the wings of the bomber rolled him right over into the fuselage of the bomber. He couldn't control it and the whole works went in. It was terrible. The government had one Valkyrie left and they put it in the United States Air Force Museum in Dayton, Ohio. Instead of having a supersonic airplane carrying bombs, the Air Force thought missiles would do the job. They scrapped the program.

They didn't know about vortices in those days. We used to call it prop wash. If you got in behind a big plane with a small plane there was wash from the propeller and it would whirl around out behind the airplane. If you flew into it with a little airplane you'd roll over. After the war the Air Force ran tests with smoke and they found out vortices were created off the wingtips. When a big airplane came in for a landing, as it slowed up the vortices got bigger. They went straight down the runway behind the plane. If the wind was out of the right, the vortices would move a little bit to the left. You had to land accordingly to avoid it.

I landed at Logan one time, it was very quiet air, smooth and beautiful. I was in my twin-engine airplane and I got caught in the vortices of a big jet in front of me. I didn't see it, and as I was landing I rolled up and the tower saw me. They started to yell, 'Are you all right?' I tucked the nose down and got out of the vortices to the right and then came and landed. The bigger the airplane, the bigger the vortices. A little airplane caught in a big plane's vortices is thrown around like a toy.

Reno Army Air Base is now used for races. There were several training bases like it all over the country. The United States was full of airplanes. England was like one big airport. All it had was field after field of airplanes, getting ready for the Invasion.

After Reno I went home to New Castle Army Air Base. I arrived there and was called in to the squadron commander's office. I was in the 89th Ferrying Squadron. Jack Daunt, my commander, said, 'I've got some good news for you. You got a promotion.' I said, 'Why? I'm a command 2nd Lieutenant.' The command pilot was the highest pilot.

Kidding amongst ourselves, if we were in a grade too long we'd call ourselves command 2nd Lieutenant or command 1st Lieutenant. He said, 'Well, you got the highest rating in Reno of anyone to graduate. You got a superior.'

When I first arrived at Wilmington, Captain Mitchell gave me a rating of good on my efficiency report. The guy gave me a good within a week and he didn't even know me. I said, 'What does that have to do with it?' He said, 'From that day on you got excellent in everything. Every airplane you flew, every flight leader you had. But you had one mark against you of good. You got a superior in Reno and that overrode the good. Your average is excellent and so you get a promotion.' I said, 'Well, if you thought that I was that good and I did all that flying for you, why didn't you give me a superior?' He said, 'We don't give superiors.' I was a 1st Lieutenant after 18 months as 2nd Lieutenant. Everybody else was getting promotions at six months except for me.

The very next day I got a trip to go to Buffalo and pick up a C-46. They assigned a copilot to me who was a sergeant. I got to Buffalo, checked out the C-46 in the factory and I went taxiing out. I heard someone talking on the air. It was Al Lechshied, who I had gone to join the Air Force with years before. We talked as I sat on the ground. He was flying test operations in a fighter plane, a P-40. He was a test pilot for Curtis. He flew for Northeast until the war ended and he went to work flying a C-60 for John Fox. Fox owned the *Boston Post* and was the president of Western Union. I talked to Al for a while then I took off and headed to my base in Delaware."

CHAPTER FOUR

The Mission:

"**I WAS SENT TO FT. WAYNE,** Indiana, to pick up a C47. I flew it back to my base. My commander said I was to fly the plane to England by the southern route. I said, 'How do I get there?' He said, 'Well, these are your orders. From one base to another they'll give you instructions.' The route was from Wilmington, Delaware, to Homestead, Florida, to Borinquen Field, Puerto Rico. From Borinquen I was to go to Atkinson Field in British Guiana, now called Guyana. Then I had to go to a field in Belém in northern Brazil. From there I was to go to Natal on the bulge of Brazil. Then I was to go to Ascension Island in the middle of the South Atlantic. Next I would fly to Africa and land at Robert's Field in Liberia. From there I would go north to Dakar, Senegal, on the west coast of Africa. Then I was to fly over the Sahara Desert and the Atlas Mountains into Marrakech, Morocco, in North Africa. From Marrakech I was to go to England.

This route was much longer but there were so many airplanes in the world moving to England for the invasion that there was no room in the north for anybody. The northern route was filled to capacity. If bad weather was in one place it would plug everything back up into the States. We didn't have range to go directly across into England. The planes didn't hold enough fuel.

NEWCASTLE ARMY AIR BASE, FEBRUARY 8, 1942. SECOND LIEUTENANT JULIUS GOLDMAN (BOTTOM RIGHT) PREPARED FOR DUTY WITH OTHER OFFICERS IN HIS SQUADRON.

I took off from my base in Wilmington in a C-47. It had four extra barrels of gas tied into the fuel system with 100 gallons each. All totaled, I had 1,204 gallons of fuel which would carry me ten or twelve hours. On a long range cruise, I could get fifteen hours out of it. I was assigned a crew that I'd never seen before. Arthur Lamb was my copilot. My navigator was Paul Rajcock. My radio operator was Francis Sheehy and my crew chief was Jerome Elgarten, a kid from the Bronx. Rajcock had one trip under his belt, my copilot Lamb had no trips, the radio operator had two trips and my crew chief had never been in an airplane before. He had just been through school. I followed the flight plan down to Florida, and my navigator followed his flight plan, too. He estimated seven hours, 30 minutes. I landed in seven hours and 32 minutes. That gave me some confidence in the navigator. We had some weather to fly through but everything worked out perfectly.

We landed in Homestead, Florida. I had to go through a weather briefing, medical briefing and route briefing at each stop. I said to the crew chief, 'When we get down to Puerto Rico, I want you to check this airplane thoroughly. In South America, I want you to make sure we're all set before we jump across the ocean.' Crew chiefs were responsible for changing the oil and filters and checking the engines. We took off that night at midnight for Borinquen Field. My navigator said, 'The first guys out get in first. When you get down to Borinquen, it's going to be a zoo.' He was right. There were sixty, seventy, eighty airplanes leaving at a time, one

after another. There was no control in between. Once you were cleared off the base you were on your own.

We were the first ones out of Borinquen, Puerto Rico. We got in the air and I had trouble with the radio. It had an ADF radio that was hooked to the inverter. The fluorescent lights in the cockpit were hooked to that inverter. I couldn't switch them on, so I made an instrument takeoff with white lights. It was foggy and hazy that night. I hadn't flown this airplane before, it was a new C-47. I wondered how I would find this place if I had trouble.

I heard somebody yelling, 'I'm having trouble, I haven't got enough power.' It was Schmidt. When the pilot is busy on the takeoff the copilot pushes the throttles to the correct manifold pressure. The kid he had for a copilot didn't know that, so when he took his hand off the throttles they slid back. Schmidt was at 50 percent power and he was coming to the end of the field. He wasn't getting any air. Just in time, he reached forward and slammed the throttles. He got into the air just in time.

I was off the coast of Venezuela, flying at 9000 feet in and out of clouds. We flew through a tropical front. It was full of tremendous thunderstorms. Flying into one of those thunderheads can be very dangerous. It was daylight. Whenever I'd see a thunderhead I'd work around it, I would never enter it. A B-25 went by me, at a higher altitude. I saw him enter a thunderhead and I didn't see him anymore. I kept looking and then I saw him at 12,000 feet, right above me. The thunderhead had sucked him up and threw him out, like a pea spit out of a pod. He was in one piece and he was way the hell up there. The weather might have bent the airplane it was so rough. I thought, 'He'll never do that again.'

Eventually we got off the coast of Venezuela, we got out of the clouds and I looked down and saw an island with a blimp on it. I asked my navigator, Paul, 'What is that on your charts?' He said, 'Those are little blimps they use to scout for submarines.' I said, 'Are there any big hills around here?' He said, 'Once you get beyond here, there are no hills.' I said, 'Next time I see a hole, I'll spiral down underneath and we'll go right into Atkinson.' He said that was a good idea.

About thirty minutes later I descended into a hole, and I saw the jungle. I leveled out at about 400 feet. I could see pretty well underneath the clouds. I was going over villages and I'd practically blow them down I got so close. All of a sudden I flew into a rain shower that was so heavy I couldn't even see the engines. The water was leaking into the cockpit. I came out of that shower three minutes later. I saw a hill higher than me to the right. I said, 'Paul, you told me there were no hills.' He said, 'The charts aren't accurate.' I started to pull back to go up in the clouds and I figured I better not do that because I wasn't sure where I was.

Just then my crew chief came up with a cup of bouillon and put it on the pedestal for me. I was too busy flying to drink it. The cup fell off the pedestal into my chukkerboot. I had a hot foot and I was flying in heavy rain and I couldn't see anything. We eventually got close to Atkinson Field, British Guiana. On my ADF I was homing on the station. I called them and they said, 'Be careful. There's a B-26 coming down through the clouds making an approach.' The man in the tower said, 'You'll be the second one to land, you can't be the first one.' I stayed wide of the station on my ADF. As I turned and came back towards the station the B-26 came right down out of the clouds in front of me at 700 feet. I went in right behind him. We landed on the field in the middle of the jungle.

EACH STATION ASKED ME to take mail. I'd take on 500 pounds here, 300 pounds there. We stayed overnight, picked up the mail and got fuel. We took off once again at midnight and got into Belém. It was an easy ride. I told the crew chief to be sure to fill the barrels with fuel. Elgarten said, 'Yes sir,' and I went off to the briefings again. Up until then, I had only filled the regular tanks. I filled the barrels up to see how much fuel I was burning per hour on the long range cruise between Belém and Natal, Brazil.

When we left Borinquen my navigator had a box with twelve pigeons in it. I said, 'What's that for?' He said, 'If we go down there's no way they'll ever find us. I can put the coordinates on the leg of each one of these pigeons and they'll home.' We changed pigeons in Atkinson, and in Belém. With the pigeons, help would come and look for us at the coordinates. I thought that was a good idea. Otherwise no one would ever find us.

Something was bothering me. I wanted to go to my airplane. We were getting into South America where

OPPOSITE: JULIE
BECAME A
FLIGHT LEADER
AGAIN ON HIS
TRIP TO
ENGLAND,
LEADING YOUNG
PILOTS HEADING
TO ITALY.

there was more daylight. It was in the spring, maybe March. I went out to the airplanes and gave the colors and security codes of the day to the guards. My airplane was in with over 150 other planes. I saw a clipboard hanging in a wheel well with a paper on it. It said, '*Could not fuel the airplane as it was locked.*' The crew chief had locked the airplane and went over to the other side of the base. I went berserk. They told me he had to be with the enlisted men. I got a wagon to take me there. I found him shooting craps, and gambling. I reamed the hell out of him. I said, 'You little bastard, are you trying to kill us? You get out of this game. Go back and get that airplane filled.'

I went to sleep and got up early. It was mostly a daylight flight. I got to the airplane and the doors were locked. The note was still there and Elgarten was sitting on a step. I said, 'What are you doing here?' He said, 'Because we're packed so tight in the airplanes, no one could back up the semi-trailer tanker. I couldn't get the gas.' I wanted to kill him, I said, 'You're going to stop me from getting out now with all this fair weather.' I said, 'Where's the truck?' He showed me where the trucks were parked. I was used to driving a semi-trailer in our trucking business and I was able to back it up through the airplanes. I filled up the barrels and pulled the truck out. We got our flight plan and took off.

I figured out what my consumption would be per hour. I knew I'd have no problem crossing the ocean. I landed in Natal on a nice sunny day. I went into operations. I said, 'I had a 50-hour check on the airplane

engines on the way down. I want to get some more checks.' They looked the engines over and said, 'Nobody did anything to the airplane.' I said, 'What are you talking about? My crew chief was supposed to pull a 50-hour on the way down from Puerto Rico.' I got a hold of the crew chief. He said, 'It was raining all the time, the weather was bad,' and he hadn't done anything.

I blew my cork. I went to operations and said, 'I don't want a crew chief. I'll carry more mail, anything, I don't want him.' The officer there said, 'You can't go without a crew chief.' I said, 'He's useless.' The man said, 'We'll put two men on one engine. He'll go on another engine and we'll do the 50-hour.' Later, I went out to the airplane. They were buttoning up the filter cover backwards, with Elgarten. I screamed and yelled at the officer in charge of maintenance. 'You've got to go out there and stand there and watch them. They're going to kill me!' The officer had everything rechecked and turned the cover around so we were all set.

I went to the route briefing. The commander said, 'Goldman?' I said, 'Yes sir,' I stood up. He said, 'You're a flight leader.' I said, 'How come? How can I be a flight leader, I can't even take care of myself. I got a crew that doesn't know what the hell they're doing, I don't know what I'm doing.' He said, 'You know how much time these kids have? There are kids here that have no more than 300 hours total time going to combat.' At that time I had 1800 hours. He said, 'You've got to take it, how do you expect one of these kids to be a flight leader?' I said okay. We were set to take off at midnight.

I was in charge of five planes, four others and myself. We talked to each other on the radio. It was raining like hell and lightning was flashing. The runway was lit up with flarepots—kerosene lamps that look like basketballs. The kids took off first, I followed them out. I got half way down the field and I could see what I was going into, and I thought those kids were going to get killed. In snapping lightening, heavy rain, and turbulence, it took me an hour and 40 minutes to get to 11,000 feet—twice as long as usual. I couldn't see them. They were supposed to be in my flight, and we were all in the weather.

In my climb, there was a flash of lightning. I saw a big black cloud right in front of me. I rolled the airplane over to go around it. I looked down and saw another airplane under me, one of my kids, doing the same thing. I was about 1000 feet above him. He made a left turn and went around the cloud. Once I got to 11,000 feet, it was clear. The moon and the stars were out. It was smooth so I put it on auto pilot, and got everything squared away. The engines cooled down—they had been running hot.

I couldn't see all the other guys but we were talking to each other on the radio. If they were within 50 miles of us we called the flight. We called operations at Natal and told them we were all okay. It was about 1500 miles to Ascension Island. My call sign was Typeset Man. They gave us different code names so that the enemy couldn't tell who we were. Submarines monitored all of this and they'd shoot at planes from the surface. The navigator would talk to the other pilots just to check on them but otherwise would keep silent on the radio.

I got a call from Cohen, a Captain of one of the DC-3's. He was in the clear, and he was taking a risk because he should have been using code. Cohen said, 'The engines are giving me a terrible time. They're rolling and rocking and overheating. My fuel consumption is up double what it should be.' He said, 'I've got my mixture rich because I'm trying to cool the engines down.' I said, 'Well, let's relax a little, and let's go over your checklist.'

We had checklists that were about two feet long. I started off the checklist with him, and I said, 'What position are your carburetor levers at?' He said, 'They're all the way up.' I said, 'They're on full hot. That's what your problem is. Lean them right down to cold.' I said, 'Close your cowl flaps to the trail setting. As soon as the engines cool down, stabilize your mixture.' He called me back about 45 minutes later and said he was going to make it. The engines smoothed out and the fuel consumption had dropped.

WE WERE MOVING along getting close to Ascension and I had to go to the john. I told my copilot, Arthur Lamb, to switch the tanks in fifteen minutes. He was reading a Wild West story. He said, 'I'll take care of everything.' We had switched from the main tanks to the barrels on takeoff. We were burning the 800 gallons, and then we needed to go back to the main tanks again.

The crew chief was right behind him in a little alcove. He was sleeping in there. I shook him. I said, 'Elgarten, remind Lamb to switch the tanks in fifteen minutes.' He said, 'Yes sir.' I walked through the plane. My navigator, Paul, asked if I wanted to learn how to shoot a navigational sunline. I knew nothing about celestial navigation. I said I would do it when I came back. I went by the radio operator and talked to him and then went in the john. I came out and talked with Paul. I was standing on a bench under the astrodome lining up the sextant. As I pulled the trigger to shoot the sunline, both engines quit.

I jumped off the bench, and dove through the hole into the cockpit. I looked at the instruments and saw the fuel pressure was down. I struggled into my seat. Meanwhile, Paul was trying to hang a Mae West life vest on me. Everybody else was already wearing one. He was choking me trying to get it on because we were going down. First I put on the main fuel tanks. I ran the wobble pump and yelled to Elgarten, 'Switch off the barrels, switch off the barrels.' He was stunned, he didn't do anything.

Once I got the main tanks on and got the pump going, I had a little of the fuel. The engine would start up and then quit, but I was descending all the time. I was at 11,000 feet before the engines quit. With the barrels shut off and the main tanks on. I got the wobble pump going. I built up pressure, and the props were spinning all the time but there was no power. When the engines quit, the plane was on automatic pilot trying to hold itself level. The props were milling with no power. I shut off the autopilot and then she started right down.

If I left the plane on autopilot too long she would have stalled. I didn't get the engine back under full control until I got down to 4000 feet. I got everything squared away. Then I reamed Elgarten and then I reamed Lamb. It was Lamb's fault. He just kept reading his book. Time was nothing to him. From then on, I didn't trust him. I wouldn't trust anyone. Everything I did myself, I learned I must do it myself. Everything that had to be checked, I checked and rechecked it. Before this happened I would give Lamb a landing or a takeoff. I cut him off completely. I let him sit there and hand paper to me when I needed it. I didn't talk to Elgarten either.

We went into Ascension Island, which was controlled by the British. On the island is a volcanic peak. The military had cut the peak off and made a runway on top of it. They cut the center hill out and made a runway at about 4500 feet. We stayed overnight and got refueled. We took off and went to Robert's Field in Liberia. It was run by Pan American at that time. Pan American was the commercial airline that flew around the world most often. They built Robert's Field for the United States government and took care of it. It was also a Firestone plantation area, where they grew rubber trees. We got squared away there and took off for Dakar, Senegal.

We landed in Dakar on steel mats. It was on the edge of the Sahara desert. The steel mats interlocked with each other so the planes didn't sink in the sand. We decided to keep on going because the wind was blowing dust all the time. We headed out over the Sahara. All the kids were in trail. We couldn't see them but we were talking to them. We kept climbing in a dust storm. We flew at 10,000 feet for three hours and we never topped it. I was worrying about the engines getting so much sand in them. I thought they would seize up but they were fine. I found a pass through the Atlas Mountains and descended. All the kids followed me down. We landed in Marrakech, Morocco, at a big field with 400 airplanes on it. That's where I saw my first B-29.

Marrakech is where we separated. The kids were going on to Italy. The following day the deadline for takeoff was 3 p.m.. We were going to have B-24s protect us against the Germans that were in Spain. Out of about 200 airplanes, I was the last guy for takeoff. Most of the B-24s and bombers left ahead of me, and I had no cover.

I got into the air and climbed up to 9000 feet. I got off the coast of Portugal. The clouds closed in from the top and bottom, and now I was in weather. Just before it closed in I saw one of the C-47s that was going to England returning on one engine. He went into Lisbon, Portugal, a neutral country, where we had ambassadors that took care of him. I didn't know what happened to the others, I couldn't talk to anyone because we were in the clear and there was silence on every channel.

The rain started to freeze onto my plane. On the wing were boots. Like a shoelace, the boot wrapped around the leading edge of the wing. The boots filled up with hot air and broke off the ice. There were also boots on the vertical fin and the stabilizers in back. I couldn't run them constantly. Once the ice was gone I'd turn them off. I would let the ice build up on the ribs again and then turn them back on. I had 15 gallons of prop de-ice which went out to the propellers on both

engines. I put that on slow drip so I would have enough for the trip.

This airplane had a new gadget, a radio altimeter. It measured how high I was off the water or the ground. I had fooled around with it going across the hills on my trip to Indiana. If it was red I was too close to the ground. If it was yellow I was getting close, and if it was green I had good clearance. I put on the radio altimeter but I was way up in the clouds and it didn't work. The radio altimeter had long, pointed antennas that stuck out. Ice built up a foot long on the antennas and then it would break off. The nose domes of the prop would build up with eight or ten inches of ice. It would break off or just stay there spinning. I had auto pilot on the plane but I had to hand-fly it because of the ice.

After a while it started to ice up so much on the leading edge that I decided to descend. There was a ball of ice around the air speed and the heat didn't reach it. I was flying without an air speed reading. After about four hours of flying with the ice, I cut the nose down to what I thought was the right angle. I descended. I knew if I went too slow she would stall and spin in. I got down to 1500 feet. The ice started coming off the airplane in chunks everywhere. At 1000 feet I stabilized it and I had my air speed reading back. I didn't need to use the prop wash or the boots any longer because it was just raining.

I got another 100 feet down and I could see the water. I was flying across the waves headed for England. Navigation was done by Morse code. I flew by listening to the beeps. I was flying the 10th meridian at 1000 feet. I saw a convoy in the water and a rocky coast ahead. The convoy didn't know who we were and I thought they might start shooting at us. I pulled up into the weather. I said to my navigator, 'Where are we?' He couldn't help me because there was nothing that he could see. I said 'Well, we must be off the coast of France.'

My radio man heard someone say Gloucester, England. He thought we were over Ireland. I was over the water, so I knew Ireland must've been the rocks I flew by. I turned around and went back, flying at about 700 feet. I avoided the convoy and came across the rocks again. There were big boulders standing up 1000 feet in the air. I kept working my way around the rocks. I saw an oval marker on the ground. It read, 'Eire 29.' I looked in my book and saw my location.

I started to fly by the numbers around the coast. I couldn't go over Ireland because they were neutral and there was a possibility, not knowing who we were, they could shoot us down. They had large balloons in the air to deter planes from entering their air space. An operator from Gloucester, England, called my radio operator. We gave them a steer as to where we were.

The Royal Air Force had two Mosquitos. They were very fast twin-engine wooden fighter bombers. They were going to come out and pick me up. I didn't trust them. At my briefing, we were told the Germans could have another homer with the same frequency on a station in Spain or France. They could steer the pilots right to the enemy. I just kept flying by the numbers. They never picked me up because I kept going along at a low altitude. I flew all the way to the Eire 14 marker. Then I headed directly across to Valley Wales, England.

The English operator kept calling me but I didn't trust him. My code name for the trip was Bluejay but they never called me that. I followed my automatic direction finder, and I tuned in the frequency. I landed in Valley Wales 12½ hours after I took off. It was a hard flight and everyone was glad to get on the ground. Gloucester called me again after I landed. In his English accent, the operator said, 'If you don't do what we tell you next time, we'll shoot you down.'

The crew and I were going to get something to eat. We hadn't eaten in about 15 hours except for a cup of tea. The base staff drove us in a six by six truck, where the soldiers faced each other. We were bouncing along a dirt road. The truck had open rips in the back. We got to the mess hall, about five miles away. I went in with my two officers, Lamb and Rajcock. The enlisted men ate in a different place. My food cost 68 cents a meal, it was very cheap. I put my hand in my pocket to get my money and my wallet was gone. I'd lost it. I went wild. I jumped in the air, dropped my food, ran out the door and went to the truck. I said to the driver, 'I'll give you $10 if you drive me back to the airplane. I must've left my wallet in the airplane.'

JULIE AND HIS CREW FLEW BY WAY OF SOUTH AMERICA TO AFRICA TO GET TO ENGLAND IN A C-47. THE NORTHERN ROUTE WAS CLOSED TO ANY MORE TRAFFIC.

When I left my base in Delaware, my crew suggested that we wait until the quartermaster opened so we could get our pay. I wanted to take off at 7 a.m., but it didn't open until nine. I took a vote and everyone said wait. I went to finance and I got my pay which was about $300 all together. The crew got their money. I got a second envelope with $2500 in it. That was $500 a man. If I was forced down in neutral territory, we could negotiate and buy ourselves out. The Air Force didn't want to lose the airplane and us, so I had $500 a man for a five-man crew. I gave $500 to each one of my two officers and let them hold it. In case I lost the money, I would have the other $1500 (my two enlisted men's money plus mine) and my $300.

A lot of guys were gambling with the money and saying they lost it. Before I took the money, I was told if I lost it I would have to pay for it. I could see myself paying $1500 plus my own $300. We were charging down this dirt road and I was looking and thinking, 'I'm in trouble.' Then I saw a brown spot in the road. We slid to a stop. It was on the other side of the truck. I went around and there was my wallet, with all the money in it.

THAT NIGHT AT THE BASE in Valley Wales, I thought of my cousin Sonny. Sonny was drafted into the military, and he was blind in one eye. His mother had given me a box of goodies and little cans of stuff. I thought if I could find him, I would take it to him. I went to our intelligence and said I wanted to reach Sonny's base. They said they couldn't do that, for secrecy reasons. I went to British Intelligence and asked them,

and sure enough they found the base. I called the base and asked if anyone knew Sidney Goldman, which was his real name. The man who answered was his commander, and he said no. While we were talking there was a sergeant who said, 'Hey, we have a Sidney Goldman in the next tent.' I talked to my cousin and he asked me to come over. I couldn't get round-trip transportation. I'd try to visit when I came back the next time.

My officers and I were put on a train in first class. My two enlisted men sat elsewhere. It was an oil train. They had chains that hooked the cars together with bumpers. We headed out to Prestwick, Scotland. We were clacking along on the train and they stopped us in a town to get us something to eat. They gave us fish cakes that smelled bad, and the coffee was cold. I threw the fishcakes into the railroad tracks when no one was looking. I went back in and tried to sleep sitting up with my feet on the bench across from me. That's the way we rode all the way to Scotland.

It took all night to get there. In Scotland they told us we would fly out on a four-engine airplane from Prestwick the next day. It was very cold that night and they put us in a Nissin hut. It had no ends in it, and there was no heat. I got some blocks of coal they had dug out of their mines and tried to make a fire. We couldn't make a fire with it. We had heavy clothes but we froze all night.

Finally we got on the airplane. My crew chief and the others were playing cards. They said, 'Why don't you come and gamble?' I said, 'I don't gamble.' I could see now why they said if I lost the money I had to pay for it. They finally financed me. I won about $100 from them but I gave it back. I never gambled in my life. I never drank and I never smoked. We got back to the States and I wrote up a hell of a bad report on that crew chief. He wanted to become corporal and he asked me to recommend him. I said, 'I don't think so.' He was no good. He was useless.

PILOTS HAD TO HAVE a medical every six months. I came back from Scotland and was given a medical. I had albumin in my urine so they sent me to a military hospital in Valley Forge, Pennsylvania. I stayed there four or five days. I said to the nurse, 'I don't want to stay here. These guys are all wounded.' They all had bad burns from airplane crashes and other injuries. I said, 'I'd like to go home.' She gave me back my uniform and I went home. She called me when she knew the doctor was going to come in. They didn't have enough doctors to keep an eye on everyone. I was home about a week, maybe ten days. Nobody knew where I was except her. She called me up one night and said the doctor would be in at 9 a.m. the next day. I jumped in the car with my uniform. I drove through the night to get back there in time. In those days the speed limit was 35 miles an hour. I was doing 60 or 70 all the way to get there by morning. She gave me back the hospital clothes and I jumped back in bed. She was a nice girl. I was eventually sent back to my base once tests showed no albumin.

I kept flying different airplanes all over the country. I flew fighters and C-47s. I was given orders to go to

THE TWINS, MARLENE AND MYRNA, WITH THEIR FATHER ON ONE OF HIS REGULAR STOPS HOME DURING THE WAR.

Baltimore and pick up dive bombers. Some A-25 planes had a lot of trouble with their vertical fins so the Air Force wasn't using them. They were sitting in a part of the airport right along the railroad. A senator traveled by there to Washington every other day. He saw all these planes sitting and he started to complain. The money for these aircraft was given by cities who advertised their donation on the side of the plane. The senator complained to the War Department. The War Department wanted them moved immediately. I was sent to Baltimore to fly them to Aberdeen Recruiting Ground.

At Baltimore airport, I saw a man working on a P-51. It was Harry Lerman, my wife's brother. I said, 'Harry, what are you doing here? I thought you were on B-24s.' He said he was a gunner on a B-24. At high altitude he had a cold, and the pressure cracked an eardrum. They made him a mechanic. I said, 'I'll be back tonight with a UC-78. I'll pick you up and take you to Boston to see the family.' He was so unhappy that he wouldn't go with me. I went to Boston with Paul Rajcock and had Chinese food with Florence. I left there about midnight and got back to my base. I was allowed to use the airplane as long as I put in five hours of flying for them. It was over five hours to go from Wilmington to Boston. I continued to transport these aircraft back and forth from Baltimore to Aberdeen. Then I got a telegram at Baltimore telling me to come right back to the base.

THE TELEGRAM SAID Colonel Bohl wanted to talk to me. I flew back and went to headquarters. Colonel Bohl was the head of Crescent Airlines. It was a four-engine outfit flying over the top of the world into Europe and all the way into India. The colonel had been looking through the records of people for an assistant chief pilot. He spotted my record out of Reno with the superior grade. He wanted me to go to Newfoundland and check out on C-54s and be the assistant chief pilot. C-54s had four Pratt & Whitney engines. Each engine had 1450 horsepower. I said, 'Well, you can't do this. I'm frozen on a special order by the commander of the air transport command, General Noland in Cincinnati.' I was to go to the Hump.

He said, 'I can have you if I want you. Do you want to come?' It was the furthest thing from my mind that I could ever fly a four-engine. I had 4-P and by flying four-engine I got the highest rating in the Air Force, a 5P. I leapt at the chance. I said, 'Of course I want to go.' He picked up the phone and called General Noland. He had a long conversation with him and Bohl said I'd be getting orders.

Off I went back to Baltimore flying the dive bombers to Aberdeen. While there, I got a call to pick up a C-47 in Ft. Wayne, Indiana, and take it back to my base in Wilmington. I had a copilot who had been over in Italy and was a spotter on a C-47, dropped the paratroopers and the whole squadron was shot down except him. He became a basket case. He claimed that some of the airplanes were shot down by our own guns. They had shipped him home, and he became my copilot. He was a very good C-47 copilot.

I was in Wilmington about four days when I got new orders. I was to take this C-46 that I had picked up in Ft. Wayne, Indiana, and clear for the Hump. I guessed that Colonel Bohl didn't have the power that he thought he had. I wasn't afraid to go to the Hump, I never even gave it any thought. My friends were there. They had lost so many airplanes flying into the Hump. C-46s were big twin-engine airplanes. If they got in any kind of trouble, they couldn't get over the Himalayas and they would crash. They needed pilots desperately and that's what I had been trained for in Reno. It was a tough place, but it didn't bother me.

In order to clear the base I had to go every place where I had done business and fill out a form. I had to turn in my parachute and equipment, and get signed out

on my insurance and my wills and trusts. It took two days to cover every station on the base. The last station was the headquarters for operations. As I walked in, the operation crew was standing in the doorway with Colonel Bohl. He said, 'Hey Goldman, what are you doing?' I said, 'I just cleared the base. I'm going to the Hump.' He said, 'Impossible.' He jumped up, turned red and said, 'Come to my office.'

I followed him into his office. He picked up the phone and called General Noland. He shouted into the phone, 'You told me I could have anybody I needed on this Crescent Airline of ours. I told you I needed this man and you okay'd it.' There were some words between them. Bohl said VOCO, which meant verbal orders of the commanding officer of the air transport command. Now he didn't need written orders. He hung up. He said, 'You finish up what you're doing. Tonight you're going to be on an airplane headed for Newfoundland.' I said, 'I haven't checked out yet.' He said, 'You'll be checked out on the way there and you'll get the balance of it in Newfoundland.'

I had to check out on a C-54, a four-engine plane. I went out and sold my car to the first guy who wanted it, for $600. I was sent to Newfoundland and got my

HARMON FIELD, NEWFOUNDLAND SERVICED C-54S. THE PLANES WERE FLOWN CONSTANTLY. ONLY THE CREWS WERE RESTED.

THE AIR TRAFFIC
CONTROLLERS
HAD THEIR
HANDS FULL.
MORSE CODE WAS
USED TO KEEP
TRACK OF WHERE
PLANES WERE
FLYING.

training up there on four-engine planes.

Crews waited in Newfoundland. The airplane would land with a five-man crew. A five-man crew would be rested, and waiting. That crew would take that airplane on to the next stop. They'd fly from Newfoundland to France, or the Azores, into Egypt. They'd fly west until they got to the Himalayas and then they'd fly over them. Airplanes were coming back from these places as well, and there'd be a crew waiting to take that plane back to the States. The crew would rest for a few days and then they'd start out again. Every station had 12 or 15 crews sitting there waiting. The airplanes wouldn't wait. They'd fuel them, take care of them and go on.

My job was to take care of the crews and make sure that they were sober. A lot of guys went to the club and had a few extra snorts, or shots of booze. If they were smelly, I would take them off and put another crew on. I never let them on unless they were perfectly sober. The rule was 12 hours from the bottle to the throttle. I didn't drink so I could smell the stuff. We had to pull crews off and they'd get mad. I wouldn't let them out and take a chance. They'd try to fool you and cover the smell but I could still smell it. This lasted for about six months. Then the chief pilot was sent back to the States. From Assistant Chief Pilot they made me Chief Pilot. I became the test pilot for the four-engine airplanes there. Anytime they had trouble with an engine, I'd fly it out.

I got to fly other aircraft. I flew PBY's, huge twin-engine flying boats. I flew a Norseman. It was a single-engine, tail dragger built in Montreal. We had a Cub on floats and a C-45 Twin-Beech. I flew all different airplanes.

They had a call for a person to fly a Grumman Duck back to Presque Isle. It was a tough airplane to fly, an amphibian biplane with a 975 horsepower Cyclone engine in it. It was covered with dust. The base commander, Colonel Morgan, was being transferred from Stephenville to Presque Isle and he wanted the airplane there, because he used to fly it. They looked in the records and thought because I used to fly Pursuit, I could fly that.

That was one dog of an airplane. I got it out, checked myself out in it, took off and it made more noise than 15 P-47s. On takeoff it had shot blades and they used to turn at high speeds. They could hear me from one end of the base to the other. I got in the air, shuffled around, and tried to make a landing between snowdrifts. I hiphopped through there and didn't stop. I went through the field and came back again. I made a real low taildragger landing and got it on to the runway. I made half a dozen landings and takeoffs and then the airplane was ready for Presque Isle.

Ray Fulcher was a young guy who was an instructor for a flight school. He was drafted into the Air Corps and became my crew chief in Newfoundland. He had been an instructor and there he was gassing airplanes. I took him out with me in a Norseman and let him fly it. He did well. I had Ray promoted to be a crew chief on a flying boat, a PBY. He was so grateful for it that he made me a bracelet.

He got a piece of propeller from a crashed airplane and a bronze bearing from an engine. He made wings out of the bearing and attached it to the bracelet. I used to take him fishing with me. We'd land on the beach and get sand in the wheels. We caught hell back on base so I didn't do that anymore.

JACK SATZ was a captain who came from Wilmington, Delaware, with his airplanes, and he always bought a lot of stuff. In the States everything was rationed. Eggs, butter, bread, you name it. Even the speed limit was put to 35 mph because over that, the tires burned rubber. Newfoundland didn't ration anything. I got meat and booze and stuff like that. I told Jack he could use my room to store things he bought from the commissary. He then took off for Santa Maria Island in the Azores.

Jack landed on Santa Maria and his four-engine airplane had a bad engine. The mechanics changed the engine and that night it was ready for a test flight. He and the copilot and some engineers took off. The copilot sat in the pilot's seat. There were mountains all around the new strip. The copilot made a left turn instead of a right turn. They flew into a mountain and all got killed. Jack's stuff stayed in my room until I gave it all away. We were all very close to each other. I felt terrible about that.

THE COMMANDER of the Army Air Corps, General Hap Arnold, wanted to go fishing. Lee Wolf, a famous fisherman, was brought in to find the best fishing spots. A pilot and a crew chief flew the General and Wolf in a Grumman Goose to a lake near Belle Isle Strait. A day later the four hadn't returned. We went out searching for them in a twin-engine Beech, a C-45. I got a call from base saying that we should turn back and get a flying boat. Two people had walked out of the forest in Port Saunders, northwest of Newfoundland. It was Lee Wolf and the pilot of the airplane. The General was still on the lake with the crew chief. His Goose was stuck on a rock.

We went back to our base and picked up a PBY. We flew to Port Saunders and landed there in the bay. The wind was blowing pretty hard. As soon as we landed we had to spin around or we'd be hit by a schooner. Eventually we saw a local Newfoundlander rowing Lee Wolf and the pilot out to us. We got them into our PBY. They told us the General and crew chief were on the lake nearby. We took off and flew to the lake. Sure

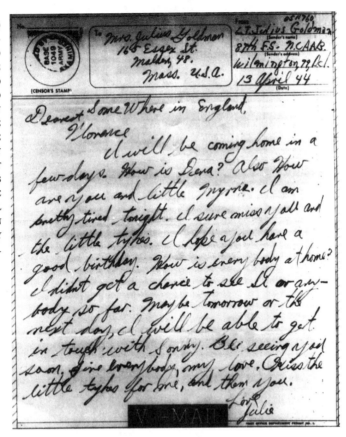

enough, the Grumman Goose was sitting on a rock. We got the General and crew chief out and flew back to the base. The General turned to me and said, 'Get my Goose out of that lake!' I didn't know how to do that.

I was sitting in the building we called the Hotel DeGink, where we could get a meal or sleep. I was talking with the others about how I had to get this airplane out. I heard a roar, 'Attention!' I turned around and looked. There stood a General with a lot of stars on him. He had a shiny helmet and pearl-handled guns. He was yelling

at the base commander, Edson, because we hadn't popped to attention. Edson was originally the aviation manager of Boston Airport. After giving Edson hell, he left. A former Pan American captain who had flown big flying boats before said, 'Lieutenant, you want to know how to get it out?' I said yes.

He had hit a rock with a Boeing 314 flying boat on the coast of Africa before the war. The plane was so valuable, they had to get it back to the States and fix it. They put burlap bags in the hole in the floor. They poured sand and cement in the hole and mixed it up in there. That cement hardened eventually and by cable they towed it off the rock. They pumped out the water, then flew it back with about three tons of cement in the belly. They got it to the States and fixed it. I said, 'Maybe we can do that.'

I requisitioned sand, cement, and bags. There was a lot of water in it. I landed in the lake and dumped sand and cement and burlap in the hole. We plugged it up and the cement hardened in a couple of days. We pulled the plane off the rock with a block and tackle, and that same captain flew it back to our base. We were circling over him and watched as he made a beautiful landing on the runway at the base. The plane started to tip forward and then she was sliding on her nose. It burned a hole in the nose and about 60 gallons of water came rushing out.

Major Barnes, Director of Operations at Harmon Field, went out to look for another place to go fishing and took our PBY. We heard his call for help on the radio. He asked the Navy in Argentine to come in with a

PBY-Catalina. Barnes also hit a rock. The Navy refused to risk one of their planes. Barnes called the Canadians in Sydney, Nova Scotia. They agreed to come and help. The Canadians flew in and they also hit a rock, but the pilot quickly managed to beach the plane. They patched the hull and flew back to Nova Scotia.

We had to go get Barnes and our PBY. The water in the lakes went down in the summer. Our PBY was on the rock and leaning over. I flew over and dropped 2x4's so Barnes could even out the plane. The crew used cement once again to patch the hole. They pulled the plane off the rock with a block and tackle. The Navy said we wouldn't be able to fly out with all the cement in there. I decided to give it a try. I started off down wind, and called for floats up. Then I turned around into the wind hoping that would give me more lift. Amazingly, I got the PBY into the air. I flew over the hills and made it back to the base.

The two incidents must have cost Uncle Sam a half million dollars, easily. But finally, General Arnold got some fish.

I HAD TO READ the letters of the enlisted men as an officer. Any letters that went home had to be seen. We weren't supposed to write down where we were or what we were doing. I would cut out the pieces of the letter with any information on it. One fellow wrote to his wife, 'You take a good look at the floor because you're only going to see the ceiling when I get home.' They wrote all kinds of stuff. I wrote letters to Florence from everywhere I was, but nobody would read my letters. We had to be careful because the letters went by ship. If a ship sank, a submarine might get a hold of the information. They didn't want anything to be said that would hurt the war effort. The phrase was "loose lips sink ships."

AN AMERICAN AIRLINES plane came in from France with a load of injured people aboard on litters. Litters were stackable cots that held injured men from the war. The copilot got very sick and he couldn't fly the plane back to the States. They didn't have another copilot available. They called operations at my base, and asked if a pilot was available to fly a C-54, a four-engine Douglas. I was in a separate group called 2nd Foreign

THE PBY-CATALINA SITS CAUGHT ON A ROCK IN BLUE MIDDLE EASTERN LAKE, NOVA SCOTIA.

Transport Group. I was assigned to the air base but I wasn't attached to it. Operations asked me if I could supply a pilot. I couldn't split up any of my crews so they said, 'Why don't you go?' I said okay.

It was the fourth of July, 1945. As I came over Boston at 10,000 feet it was snowing. I hit the line all the way down and the panel lit up as I was coming over Boston. I was right on target on the flight plan. The captain said, 'What are you doing after the war?' I said, 'I don't know.' He said, 'Well, you can come to work for American.' That was very hard to do. He said, 'You're pretty sharp, we'd like you to work for American.' I thought about it. I could just see my life on one airplane flying back and forth. I wasn't interested in that. I said, 'No, I don't want to be a truck driver.' Pilots didn't get paid a good wage then like they do now. I never attempted to go to any airline after that.

I was the only guy flying across the ocean who couldn't swim. They didn't know that in the Air Force. I never said a word to them. There were no questions when I took my private. No prerequisite said I had to know how to swim. Pa was a good swimmer but nobody ever taught me. It wasn't important because I didn't have time for it. Saul and Lillian didn't know how to swim either.

I flew to New York one day. I bummed my way back to Boston to see my wife and kids. After a short stay, I got myself back to New York. I was to check out a pilot, who was a Major. We were flying as far as Newfoundland together. He said, 'Fly us home,' so I said okay. The plane had four brand new engines in it. As soon as I got in the air, there was an explosion. The sky lit up and the #3 engine quit. We were loaded heavy. I struggled around on three engines, came back in and landed. The runway was slippery. As I was sliding down the runway, I hit a good solid bare spot. All four tires burned rubber off of them because the wheels were locked. I stopped just short of the ocean.

I told the mechanics in New York to check the engine. Later, the mechanic said, 'Captain, you must of iced up the carburetor. There's nothing wrong with it.' We lined up and took off again. We were in the same position and boom, the sky lit up and #3 quit. I said to the Major, 'Let's go on to Newfoundland on three engines. I've got a good crew up there and they'll find out what's wrong with this engine before it crosses the ocean.' He said okay.

I feathered the engine, changing the propellers to be knife-edge into the wind. As we got closer to Boston, we were flying at 10,000 feet on three engines. We weren't going too fast. I unfeathered the engine, and got #3 started in flight. I put it up to 30 inches manifold pressure and it ran alright. We got into Newfoundland. I told the crew chief, 'Find out what's wrong with the engine and fix it.' We didn't have many passengers aboard so they went on another airplane. We left the plane and went to get rested.

A crew was waiting to go to the Azores. The airplane was allowed two hours on the ground maximum. They worked on the plane and finally told me to try it out. I took it out with a copilot. I got into the clouds and boom, the same thing happened. I cut it back, shut #3 off, and we flew around for a while on three engines. I could see the carburetor gauge go up from green into red. We landed and they put on another carburetor. I went out, and boom, the same thing happened.

An American Airlines crew chief, who I knew from Boston came over. He said, 'Would you do me a favor? Follow what I tell you to do. But don't tell the military crew that I said it.' I said okay. (The military and civilian crews weren't getting along together. The military crews thought the civilians should enlist.) He said, 'The first thing you do is start at the spark plugs. It's a 14-cylinder engine so there's 28 plugs. Take out the plugs and if they're okay, go to the coils. Then check the electronic leads that go to the magneto. Then check into the supercharger.' He said, 'Tell them you want that done.'

My roommate, Chapin, was in charge of the mechanics. I asked him to check these things in that order. I went off to get a sandwich and wasn't gone 30 minutes when I got a call. Chapin said to come down, they found it. I went rushing down there. We had already held this airplane two weeks trying to get it fixed. When engines were overhauled, crew chiefs used dummy plugs and then sprayed the engine with silver paint. Dummy plugs were left in and the cylinder wasn't firing. When it was full bore, filling the cylinder with fuel, that would cause an explosion. That explosion would backfire and kill the engine. Chapin replaced the plugs and sent the plane on with a crew. This is how things happened in the military. We had to check everything twice during the war.

I STARTED A SHORT snorter in 1942. It was

OPPOSITE: JULIE CLIMBS INTO A THUNDERBOLT DURING THE WAR. HIS FLIGHT RATING BECAME A P-5 WHEN HE PASSED THE FOUR-ENGINE FLIGHT TEST.

LEFT: JULIE'S SHORT SNORTER WAS SIGNED BY FRANK SINATRA AMONG OTHERS. OPPOSITE: JULIE GOT A PHOTO OF DWIGHT D. EISENHOWER AFTER THE INVASION.

a trend in those days. Everyone kept their short snorters taped together in a jelly roll. The name short snorter came from snort, which was a small drink of booze. Anyone who flew over the Atlantic could sign one. People who shared a flight, whether crew or passenger, signed each other's short snorter. It started with one bill. We added a bill for each country we flew through. My first short snorter bill was a big American bill, which they made back during the war. I had my crew members sign it on the trip to England.

My short snorter is made out of a lot of invasion money. I have invasion money from Portugal, Brazil, Morocco, and China. The government printed this money for exchange in a foreign country. They printed bales of it. For $200 you'd need two bales of invasion

money in China. They used to deliver it in baskets. If an American had to pay for anything we'd use the invasion money. You could change the invasion money for so many lire, or francs, or pounds. When you left the country you couldn't use it anywhere else.

After I was stationed in Newfoundland, I got a lot more signatures. On a flight to the Azores, my crew chief came up. He said Amos and Andy were aboard. I said, 'I don't believe it. Give me the manifest.' Their real names were on the list, Freeman (Amos) Gosden and Charles (Andy) Correll. They were going to be entertaining U. S. troops abroad. I said, 'Bring them up here.'

Amos 'n Andy was a black-faced act on television for years. They were very famous. They came into the cockpit and we talked for a while. They had given an airplane to the government for free. I let them fly the four-engine airplane. They loved it, and they could fly a little.

Frank Sinatra, Sid Chaplain, Phil Silvers, Betty Heaton, and Faye McKensie were on the same flight from Newfoundland to the Azores. They were all going to entertain the troops as well. Frank Sinatra was a skinny little kid then. I was a fan and I liked his songs. As a new actor in the movie business, they wouldn't let him go beyond the Azores. All the girls were crazy about him and they figured the G. I.'s wouldn't like it. They took him as far as the Azores to perform and then we flew him back.

Helen Hayes was on the same flight with me when I was a passenger. She was a famous Hollywood actress. I knew she was on the flight. I said to the stewardess, 'Get Helen Hayes's short snorter for me.' I asked where she was. Ms. Hayes said, 'I'm right beside you.' I didn't even recognize her. She just looked like a little lady to me. She said she'd sign my snorter only if I signed hers. That was probably '43 or '44.

I got everyone's signatures. Faye McKensie, Frank Sinatra, Phil Silvers, Amos and Andy, Sid Chaplain, Betty Heaton, Helen Hayes and the Andrews Sisters, Patty, Maxine and LaVerne. These people were happy to see me too. I was a decent pilot. They wouldn't get killed.

I WENT TO PARIS, France, from Newfoundland. Occasionally we'd fly to the Azores depending on the winds. We'd land there for fuel and then go into Paris. Once the invasion happened, we were in Paris all

the time at Orly Field. You had to be careful where you walked in Paris. There were a lot of places with mines buried. Those trips we were flying four-engine airplanes bringing in blood and different things. We brought the wounded home. We'd have 24, 26 litters, or stretchers, in an airplane. We slid the stretchers onto racks and then strapped the person in. Sometimes we'd have one on top of the other. There were so many wounded, this way we could double the amount of people.

I was able to go anywhere on our routes checking pilots. I went from Newfoundland to England and went over to France after the invasion. I got a picture of Eisenhower at the time. The government was sending Eisenhower back to the United States, and the press was all around him. He was speaking to reporters and I was in back, trying to take the picture. I couldn't get close enough for a shot. He could see I couldn't get the picture so he had the crowd move out of the way for me.

I heard Sonny was in Paris in an orphanage in Villa Cablais. The building was empty and they housed soldiers in there. When I got into Orly Field in Paris, they put me up in the officer's quarters. I asked them for a jeep to go and try to find Sonny. They said, 'You don't rate a jeep.' They didn't have enough of them. I said I had to find a ride somehow.

Operations had a French couple drive their car around for the officers when they needed backup transportation. The trunk of the car had an open section in it. They had a chimney in it with a coal stove attached. That car ran on charcoal. They had no gasoline. They made their own fuel. The base assigned that car to me. I had some steaks, cigarettes, booze and a little box of goodies that Sonny's mother had given me. It was mainly canned goods and treats. Whenever I went into Paris I brought presents with me. I got in the car with my stuff and the two French people didn't speak English so I kept saying Villa Cablais while they drove me around the outskirts of Paris. We couldn't find the villa, and I thought they were looking at my stuff. The food and cigarettes and booze was worth its weight in gold.

I didn't take my gun with me—I left it in my quarters. I thought they might try to knock me off just for the food. Finally we wound up at an old red brick building and nobody was there. I asked some people outside if it was Villa Cablais, but got no answers. I kept my eye on the car. I stood there and looked at the building. There was a big door there but it was closed. A big army truck, a 6x6, pulled up and there were a bunch of guys in the back. It was full of American G. I.'s. I walked over and asked 'Does anybody know a Sidney Goldman here? Is this an orphanage?' Someone came out of the truck and flattened me. It was Sonny.

The next trip I routed myself straight into Paris and I found him in the city. He belonged to an M. P. outfit. I called and found out where he was through intelligence and they rang his outfit. I told him, 'I'm coming out Sonny, I have a lot of stuff for you.' He said, 'I can't leave. I got picked up for not wearing my dog tags.' Dog tags had the soldier's name and serial number and religion printed on them. If a soldier got clobbered they'd know

whether he was a Catholic or a Jew or Protestant. It was a very severe rule. He was confined for two weeks to the barracks. I said, 'You can come out for a few hours.' He said, 'No. My C. O. won't let me out.' He said, 'I'll only have to serve the time later. I'm not coming out.' I said, 'Let me talk to your C. O.'

His commanding officer was a lieutenant. I said, 'I just flew in to see my cousin and I want you to let him out.' I told him who I was. I heard him say to Sonny, 'Why didn't you tell me that your cousin was an officer.' He said, 'I'll let you out but you put your time in later.' Sonny got back on the phone. He said, 'I'm not going. I don't want to serve the time later.' I said, 'Let me talk to him again.' I said to the C. O., 'For Christ sake, I've come all this way to see him.' He said, 'Alright, he doesn't have to serve his time.'

I got a jeep and got to him. I had six thick T-bone steaks, six pounds of butter, six cartons of cigarettes, more goodies from his mother and a few other things. He got the chef to cook the steaks. The cigarettes were worth their weight in gold. Cigarettes were a nickel a pack, but they sold for $3 a pack in Paris. I was chief dispersing officer for cigarettes at my base.

In every four-engine airplane, they had a chief pilot's compartment. That was just for us. I could take what I wanted and then the rest of it would go on to the next base. That airplane went all the way to India. Each chief pilot in each station would take out what he wanted. We had 60 airplanes so a lot of stuff came through. I would take off two or three cases of cigarettes. A case had 60 cartons in it. I must of had 800 cartons of cigarettes stacked in my room at a time. I would give them out on the flight line to G. I.'s if they'd do a good job. On a holiday I'd give every G. I. a carton of cigarettes, sometimes two.

I'd give out five hundred cartons and more would come in. What I gave out, I would take. Nobody wanted Prince Albert Tobacco so I'd leave it on the plane. It would go all the way to India. I had Pall Malls and Chesterfields and Camels. No one wanted Raleighs, or Wings either. I'd just put them on the airplane and ship them off. On my table in my apartment, I had a big pile of cigarettes all the time. I had ladies cleaning my apartment. I told them to take cigarettes. They wanted the Camels and Pall Malls too. Florence smoked and I saved cigarettes for her. In those days we didn't know cigarettes were bad for our health. I'd bring her home five cartons of Pall Malls every visit. I'd get home a lot because I was the chief pilot.

I'd route myself to New York and then bum my way to Boston. I'd see her every three weeks or a month unless I was so far overseas I didn't get back. I could get on the train from New York but they were so full I often hitch-hiked to Boston. On trains, you were standing. If you had a suitcase, you sat on your suitcase. The trains were jammed with people during the war. I think I was the only homing pigeon of the Air Force. I wasn't supposed to do that. I saw Florence no less than once a month during the war, even when I was stationed overseas."

CHAPTER FIVE

The Dream:

"**THE NEWS CAME OVER** the radio that the war had ended. We all cheered. We took Very pistols and started shooting flares in the air. Everybody was shooting flares and they were going wild. It was great. Now I could go home for good.

I was taking care of my work at Harmon Field. The place didn't stop. A lot of airplanes were traveling all over the world. Eventually I got orders to go back to my base in Delaware and then I got orders to go to Westover Air Force Base in Holyoke, Massachusetts, where a lot of us were discharged. The Air Force asked me if I would stay. Because of my background they wanted me to be an instructor at West Point. I would inspect four-engine airplane cadets and get them acclimated to flying. I gave it a lot of thought. The reason I didn't do it is I knew I still hadn't gone to high school. They were bound to find out I had no education. I was at Westover and a Major was behind me. He said, 'I'm staying in, are you going to stay, Goldie?' I said no. He said, 'Why not, I'll be your general in the next war.' I said, 'God forbid.'

To be discharged, I had to have a medical and go to the dentist. I was sitting in the dentist's chair. I looked up and it was the same dentist I had when I signed in, in Manchester, New Hampshire. He was a cousin of mine by marriage. When I enlisted he said he was on his way to Europe, and now there he was. I went for the medical

JULIE AND FLORENCE TOGETHER IN REVERE, MASSACHUSETTS, AFTER THE WAR.

and the doctor said I had too much urine in my kidneys. He said I couldn't get out until it was taken care of. I said, 'No way.' I told him I was in the hospital for two weeks at Valley Forge and they found nothing. I said, 'Is there any way I can get out of it?' He made me sign a waiver and I went home.

During the war I came home on leave a few times. I kept looking at Muller Field in Revere, Massachusetts. It was all ripped up. The Army had made it into a test track for Weasels, vehicles that went through swamps. They built them in the Ford plant in Somerville, Massachusetts. They were taken on trucks to Muller Field which had been leased by the War Department.

I talked to Iz and Paul. I could have opened up the trucking business again. I had six months to validate our Interstate Commerce Commission certificate. Trucking businesses can't operate between states without it. We were licensed so after the war we could start up again, but we had sold the certificate to a company and gotten $6000 for it.

Dave Kates was a trucker who worked for us. While I was still in Newfoundland, his lawyer, Simon Cohen, found out who owned Muller Field. This took months. Simon gave me all the information and didn't even charge me. I knew who to go to and I had a jump on everybody else who was trying to buy that land. There were several different groups trying to buy it after the war. I told Florence of my plans. She knew everything.

The airport was owned by two estates, the Squire Real Estate Trust and the Winona Trust. The Squire Family went back 100 years in Revere. A real estate man named Muller married a Squire, so they named it Muller Field when it was a small airport before the war. The estates wanted $275,000 for the field. It was all torn up. The hangers all had holes in them. I kept looking to find someone who could help me financially. I became a speaker at the National Guard in Logan Airport. I talked about my experiences during the war. Richard Berenson, a fellow I had given some flight time to in the state guard, was there.

Berenson asked me what I was going to do after I got out of the military. I told him I hoped to buy that airport. I wanted to make a flight school and have a nice restaurant in front. He said, 'Let me see your prospectus. How much are you going to take for yourself?' I said, 'I'll make $75 a week.' He looked at me and said, 'You're crazy. If you don't make any money there then what do you want it for.' He said, 'Put yourself in for at least $150.' In those days $150 was a good salary for a workman. I said I was going to bring my brothers-in-law in as partners. He said, 'Don't take your family in, they'll be too much trouble. They got any money?' I said no. Berenson said to me 'If you get somebody interested, I'll take 10 percent.'

I went to a bar mitzvah with my wife and there was a fellow there that someone pointed out. They said, 'You see that fellow? He's very rich.' His name was Dr. Sagansky. I said, 'Introduce me to him.' They said, 'He won't talk to you here.' They made arrangements for me to meet him at his home in Brookline. Two weeks later I went to his home. I talked to him about what I wanted to do, and I told him I was interested in taking on a partner.

Then Doc said he'd have to discuss it with his partners. He called me up two weeks later. He said, 'I'm not interested. My partners say you're a nut. That swamp is infected with mosquitos and horseflies.' He said, 'If you get somebody interested, maybe I'll talk to you again.' I said, 'I got somebody interested for 10%.' He said, 'Who is it?' and I told him it was the Berensons. 'If the Berensons are interested, it must be good.' He said, 'Come on out.'

I went to his home and he said, 'How much money do you need?' I said, 'I need $50,000 to get started.' He told me to wait a minute. He went to another room and came out with $50,000 in bundles. I didn't even have a bag. I put the money in my shirt, and I drove home. We were living in Florence's mother's house in one room with my two twins. I said, 'Florence, I need to keep some money in here. Keep your eye on it.' She asked 'How much?' I said $50,000. She said, 'Don't leave it here.' I said it would be all right. Now we had to buy the place. They wanted $275,000 for it.

Berenson's father, Moishie Berenson, was an old real estate guy. We went to the First National Bank of Boston where the first trust was held. The other trust was at the National Shawmut. We met the fellow who was the head of the trust. We said, 'We'd like to buy the place. How much?' He said, '$275,000.' We said, 'It's too much.' He said, 'Well, you can't buy it then.' We stepped away for a

minute. Moishie Berenson said to me, 'Do you have any money on you?' I said, 'Well, I have $5000.' He said, 'Give it to me.' I said, 'What are you going to do with it?' He said, 'Give it to me.' He wrote out a note, offered $75,000 and left the $5,000 as a deposit. He left it on the desk. I said, 'How am I going to get a receipt?' He said, 'You don't have to worry about these guys. They won't steal your money.' We left.

It was a lot of money to leave there with a note. I'll be damned if they didn't take it for $75,000, as is. It was all ripped up. There was nothing left of the airport, nothing. It was just the land, and it was a swamp. I was happy as a pig in mud. I wanted to go look at my place. I went there on a November day. The government had put up a barbed wire fence all around it. I climbed the fence and jumped over. There was a lot of fog and I couldn't see much. I saw a man standing there with a clipboard, writing. He went from one hill to a hanger. I said, 'What are you doing here?' He said, 'I'm checking here for some damages, we have to make a settlement to the people who own the field.' I said, 'I own it now, you have to talk to me.' He said, 'Oh no, whoever owned it at the time it was a war asset. We do business only with the people who owned it.' I thought, 'What am I going to do?'

I went over the fence like a gazelle headed for Boston. I went to see Berenson and asked 'Is this true?' He said he didn't know. He called his uncle who was the head of the Dreyfus Properties. His uncle said that was right. The war department had to do business with the original owners at the time of the lease to the government.

I went to the First National Bank. I asked the banker if he was doing business with War Assets. He said yes. I said, 'How much did they offer you?' He said 'I'll get $2700 for both trusts.' I said, 'I'll pay you $3700. I'll give you the money now and you won't have to wait. But I want to sue in your name for myself.' He said okay, and he signed the agreement I made up. I showed the War Department assessor the damage they did to the airport and how much it was going to cost me to put the runways back in. I sued and got $37,500 from the government. I had paid $75,000 so I got half of my money back.

Next I went to the bank for a mortgage. We had to get money to build the place. Berenson and I saw a fellow named Sturgess who owned the bank, Pilgrim Trust, on the North Shore. We knew all the North Shore people were very rich. The old line families lived there—the Cabots and the Lodges. We said that we would like a mortgage. Berenson had a good background and owned liquor stores. Sturgess asked how many buildings we had on the property. We told him one building had been cut into a water tank and the other building was an old hangar. He said, 'Unless you have buildings, we don't give mortgages.'

We continued to try to get the mortgage but he said no. Then he said, 'My son is getting out of the service, maybe I'll give you a mortgage if you take my son in as a partner.' By that time I was worried. I had no money of my own. I was working with another guy's money. Now I was being asked to take in another partner. I said, 'No way, I have enough partners.' We left. About a week later Sturgess called me up and said, 'I'll give you a mortgage. Let me know exactly what you have there. How much mortgage do you want?' I didn't realize how much it was going to cost. I said $50,000, and we had a mortgage.

I wanted to build the airport and I didn't know anything about engineering. Gil Winer was a friend in Suffolk Square, Malden, who went in the contracting business with his brothers. I went to see him. 'Gil,' I said, 'I want to give you 20 percent of my operations if you build my runways for my new airport.' He said, 'I'll go see my brothers.' He talked to his brothers and they said I was nuts. They didn't want to have any part of it. Gil was interested, though. He said, 'I'll build it for you but have any engineering plans been made?' I said no. He said, 'You have to have an engineer develop plans for the runways, water, drainage and electric.' His friend, Anthony Minichello, was the head of United States Engineers for the District of Boston. He helped to build Hanscom Air Force Base. Gil said, 'I'll talk to him about engineering it for you. It won't cost you too much.'

Gil went to see Tony. Without telling me, Tony went right to work. He engineered the airport and brought in blueprints an inch thick. They detailed how deep the fill would be down below the gravel. They showed how thick the tar would be and where the electric, the slopes, and the drainage would run. He brought in all the drawings and said, 'You have to get a contractor.' I said, 'I'll go back to Gil Winer.' I showed Gil all the plans and I asked for a bid. He said, 'I'll do it good for you.' It cost $95,000, which was a good price.

While the contractors went to work, I shopped around for airplanes. I bought 12 small planes in Cincinnati. Then I got three planes in Camden, South Carolina—two BT-13s and a twin-engine UC-78. I bought a Sky Ranger, and three PT's which are trainers. I had a PT-26, a PT-19, and a PT-20. I got permission to store the planes at Boston Airport in the National Guard hangar. I landed some of them in Muller Field, the smaller planes, and put them in the hangar standing on their noses. I made down payments on planes, gave mortgage payments to the bank and paid payments on the building of the field. I needed more money from my investors. I went to my partners and Doc gave me another $50,000.

I then realized I had to get a snowplow. In Newfoundland, I saw how they cleared runways with a big twin-engine blower. I'd never seen anything like it before. The plows they used were on Ford Harrington trucks. They had four-wheel drive and big augers in the front. The plow could go through snow drifts and throw the snow 100 feet. I tried to find one somewhere. I found a war assets yard in Weston, Massachusetts. They had the same snowblower that was used in Newfoundland. I saw tags on all the equipment and I wrote all the information down. Then I went to war assets and told them I wanted to buy the plow. They had lathes, drills, sledgehammers and everything I could think to have for an airport, all of it new. War Assets told me they didn't have a plow like the one I wanted. I didn't say anything— I knew they had one.

I went back to Dick and told him I had to get a big four-wheel drive truck with a snowblower. I said, 'War Assets tells me there isn't one but I know where it is.' He said, 'Go down there and check it again.' Dick said he'd call his uncle to see if someone would go with me to the War Assets office. The head of his uncle's company knew the head of War Assets for the area, a Mr. Haggerty. I went back to Weston with three books. In each one I put down the document number and the price. I recorded all the information for the equipment I wanted to buy. Dick set up a meeting with Haggerty. He said, 'We are going to send you with Old Man Dreyfus.' Carl Dreyfus was one of the richest men in Boston. He turned up in a Rolls Royce. I got in the car and he took me to 10 Post Office Square. He must've been 90 years old. The cops bowed to him. They knew him.

We went up to the top floor of 10 Post Office Square and Mr. Haggerty came out. Dreyfus introduced me. He said, 'I want you to meet my friend, Julie Goldman.' We sat down and I told Haggerty what I wanted. Haggerty said, 'We don't have anything like that.' I said, 'Will you give it to me if you have it?' I already knew that they didn't want to give it to me. He said, 'Yeah, we'll give it to you.' I opened up my folder. I took one book and gave it to Dreyfus, and gave one to Mr. Haggerty and I kept the third for myself. I said, 'Just open to page one.' The snowblower was on page one. I had listed the make and model, document number and price, the snowblower's location and how much it cost War Assets to acquire it. Haggerty turned white. They knew they had that stuff but they were going to sell it to friends.

Bulldozers that were worth $100,000 were probably selling for $5,000. Finally, he said OK.

Dreyfus said, 'You've got this stuff. You know where it is?' I said I knew. I went down there. I paid $2,780 for the snowblower. It was probably a $30,000 snowblower in those days. I bought a lathe, grease, sledgehammers, drills, and grinding stones. I bought everything I needed, all brand new. I wanted to get the plow out of the War Assets yard immediately. I drove it from Weston to my airport. It had no plates, and no insurance on it. I was in traffic, with this huge bucket attached to the hood with swords in front of it. A guy stopped short in his car. I hit him and cut his bumper in half. I couldn't stop. It was such a heavy truck and the brakes were a little rusty. I got out and said, 'Geez, I hit your car.' I said, 'I don't think I want to go to the insurance company. I'll give you $50 for the bumper.' It was an old car so he agreed. I drove off slowly. This time I was crawling and I got to the airport.

I bought a used four-wheel drive truck from the city of Boston. The city was getting new snowplow trucks so they sold it for $3500. Then I bought a Ford truck for $600 with a small plow. We were able to stay open in all kinds of weather. My airport was open no matter what the snowstorm. Boston Airport, Hanscom Field, and Norwood Airport would all be closed, but Revere Airways was open.

We created a Revere Airways flyer to advertise the opening of our airport. Dick Berenson was worried about anti-Semitism. We were both Jewish and Dick said, 'We shouldn't have a Jewish president.' Dick listed Richard

THE FIRST BROCHURE SOLD G. I.'S AT THE SPORTSMEN'S SHOW. RIGHT: WHEN ALL OTHER AIRPORTS WERE CLOSED DUE TO SNOW, REVERE AIRWAYS STAYED OPEN.

Hoag as President. Hoag was a fellow that worked for Dick in another business. He was president in name only. I was vice president and Dick was treasurer. He wanted to handle the money. Hoag had nothing to do with the airport. It didn't make any difference to him. He was a nice fellow. Then Dick wanted to make another guy president, Tom Noore. As long as it was Noore or Hoag, he thought that we would do better.

It was a year later when I stood up and told Dick, 'I don't give a goddamn what anybody says. I want my name where it belongs.' I worked hard, I ran the whole place. I was proud of who I was and I wanted to stand up on my own two legs. Dick and I had a big argument but I became president of Revere Airways.

Before I opened the airport, the Mechanic's Building on Commonwealth Avenue in Boston was having its annual sportsman's show. It was a huge building near the Armory. Businesses rented booths to sell things and market themselves. I told Dick that I wanted a booth at the show. He said he knew people who could get us a booth. We made a composite of a little airport with buildings and runways. We advertised our opening date and said we were signing up G. I. students to learn how to fly. There were G. I.'s going through there by the scores. We signed up 1000 students in three days. Now I had a list of people to teach how to fly.

I hired 12 pilots who had heard about the airport and wanted to become instructors. I had eight instructors and four spares. I got my examiner rating back from the CAA so I could issue licenses again when the flight test was passed. Next I had to hire mechanics. My two brothers-in-law were mechanics in the service. They agreed to work for me along with a group of other mechanics.

Marion McIntire came into Revere Airways. I was her instructor at Muller Field and she became a good pilot before the war. I flew with her on her first cross-country to Chatham on Cape Cod. If you flew 25 miles or more away from your home field, that was a cross-country flight. Instructors had to teach so many hours of cross-country before the student could get a license. Marion joined the Marines in 1944 and taught navy cadets to fly. That day she came to me looking for a job. I gave her a job as an instructor. She also taught people instruments on the link trainer. She did a good job. Pilots had to pass an instrument test to get a private or commercial license. Marion McIntire eventually started her own ground school at Hanscom Field teaching people instruments.

I got a four-way radio from a fighter plane during the war. The radio weighed about 200 pounds. It had four channels, A-D. I put a crystal in it and used that in our terminal at the airport. Then I installed a radio in my plane so when I was out flying, I could contact the airport to talk to my kids.

The Lear Unicom was a two-way radio, built by Rufus Ethergart. He had a factory making the radios. Unicom used a special frequency, 122.8, the first frequency just for airports. Pilots could talk to anyone else with the same radio equipment. They started to sell them like mad. Unicom first had five channels, then ten

channels. Then they built a smaller radio with 24 channels, then 90 channels, then 360. We then knew ahead of time who was coming and going. I submitted the first application for a Unicom station in Massachusetts.

Bump Hadley interviewed me on Boston TV about all the growth our business would bring to Revere. He was a celebrity, and the first guy to be on TV in Boston. We didn't own a TV at the time. My girls went to the Slettering's house down the street to see me on TV.

CROCKER SNOW WAS AN aviation pioneer from the North Shore. One of the Wright Brothers signed his pilot's license. My airport had to be licensed on opening day. Crocker was the director of aeronautics for the State. He was a stickler for accuracy, but a real aviation professional. He flew in with his little airplane. The airport looked good and the airplanes were all lined up. Everyone was there. Crocker looked around and said, 'You've got a beautiful little place.' He signed the license, taxied down the runway and took off. We opened the airport June 7, 1946, and we started flying like mad right away.

Harold Roche was an instructor before the war. I had hired him as a pilot for Revere Airways. He said, 'I want to be the first one to hop a passenger on the new field. There are two girls in the parking lot. They'll pay $3 a piece.' I said okay. One girl got in, while the other one waited. I went to spin the prop, and I walked behind the plane. Harold didn't see a gasoline roller 100 feet down the taxi way and he ran right into it. The prop came off. It shattered, and parts of it went into the highway. The right wheel was torn off and the plane was sitting on the roller. It was a new airplane. It hadn't flown anywhere. I tried to catch the tail as he sped off and I couldn't catch it. I got to him and said, 'What did you do.' He said, 'Oh my God, I clobbered the airplane.' He had his hands over his face. We weren't open ten minutes and we had our first accident.

I helped the girl out of the plane and she left. She didn't even take her money back, she was so scared. We got the airplane into the hangar and we had to rebuild the front and the prop. I became hard on everybody after that. Nothing was done without a reason—I was an SOB. Nobody did anything that wasn't according to the book. No one who worked for me had another accident with my planes ever again.

We started flight school the day that we opened Revere Airways. We had notified 300 cadets to come and learn to fly out of the 1000 cadets we signed up at the Mechanic's Building. We were busy from then on. I was so busy giving flight tests. We had so many people, I'd be giving flight tests all day, every day if the weather was good. Then I had charters at night.

I FLEW PHOTOGRAPHERS to take pictures of hurricanes, tornadoes, fires, train wrecks, ship wrecks, and plane wrecks. They didn't have news helicopters at the time so I had little competition. We were only two miles from Boston. I made a hole in the floor of the airplane, and bolted a camera in there. The photographer had a viewfinder and he'd get right over the scene and take the pictures. The news people found that it was costing them a lot of money for people like me to do it, so they bought their own planes and equipment. I took pictures all the time I had the airport. Ollie Noonan, and Bill Hannon were photographers for several Boston papers.

Bill Hannon and I were flying once, and I said, 'We have to have oxygen.' He disagreed. We landed and the pictures didn't come out good. He said, 'You weren't over the target.' I said, 'I told you, we need oxygen.' His

BELOW: A CAMERA WAS INSTALLED IN THE FLOOR OF THE PLANE. OPPOSITE: THE ETRUSCO RAN ASHORE IN SCITUATE, MASSACHUSETTS. DANIEL SHEEHAN SHOT THE PICTURE WHILE JULIE FLEW THE PLANE.

company approved it and they bought the oxygen and put it in my airplane. Whenever we went out after that we had the oxygen masks. If we got over 12,000 feet we put them on and the pictures came out good. He needed oxygen, too, not just me.

NEIL ARMSTRONG came in to the airport. He was just a young guy, and he was involved with Civil Air Patrol. He was a college student, a good guy. When Armstrong became famous, someone told

me that I had given him a flight test. We had 1000 kids on the books.

THREE ACCIDENTS OCCURRED, all with strangers flying my planes. One man crashed on Crane's Beach. He was buzzing the beach with his wife and they were both killed. Another one went into a golf range. He was buzzing his girlfriend's house behind the golf range and he killed himself. The third accident was in Malden on a hill. A fellow from the Philippines rented an airplane. He killed himself but the passenger lived and sued me. I didn't lose any money because we were insured. The guy who sued me didn't win anything because the crash was due to pilot error. I wrote my price for insurance, that's how badly the insurance companies wanted me.

A girl flew out from Revere Airways once. She flew a Colonial Cub airplane called a Skyranger. The girl and her friend took off and it started to snow. Instead of landing at a little airport nearby, she tried to find her way back. She couldn't find us and she crashed. We knew the plane went down in Newburyport so we went looking for her. She tried to land in the highway. She hit a telephone pole and a tree and just slid down. We got to the plane but she and her passenger weren't in it. We took the airplane apart. We couldn't fly it out. We took the wings off and put it on a trailer and cleared up that mess. We found them afterwards. She and her passenger were in a restaurant having coffee. They were too scared to call.

JEANNE KAUFFMAN was a secretary who did beautiful work for Revere Airways. One day she said to me, 'Julie, they are selling grave lots down in Sharon, Massachusetts.' I said, 'Oh no, I don't want Florence to think I'm going to get killed in the flying business. She said, 'No, don't tell her.' I bought four graves for $75 a piece and I didn't have to pay cash, I paid $15 each month. We had our two daughters, Florence and myself. I paid them $15 a month. If I got killed, the lots were automatically paid off.

I WAS INSTRUCTING a student while flying a cross-country. We got out the Peabody fanmarker, where airplanes received a signal to tell them where they are. I arrived over Peabody, and I checked the weather. I was told all was okay. I called Boston on the radio and Florence Cates answered. She had been working for the CAA on route control. I hadn't talked to her since I had let her fly after Larry Hanscom died. She saw my flight plan and said 'Julie, where are you going?' I said, 'Portland, Maine.' She said, 'I don't think so. We're being covered in fog now.' They never told me anything about fog in Peabody. I was going like hell because I had a tail wind. I was almost at Newburyport when I heard the news. The fog was moving fast so I said 'I'm turning around.' When I got back, half of my field was covered in fog.

I knew she was around so I called and talked to her a couple times. She was on the Berlin Airlift as a controller. They sent controllers to Berlin to protect the city from the Russians. The Russians had blocked Berlin all around and the citizens couldn't get food or coal or anything. We put in four-engine airplanes. We had probably 500 airplanes just feeding Berlin. They had to have our controllers there to keep our guys straight. She was with the CAA, now known as the FAA.

WE BOUGHT OUR FIRST house on the west side of Malden in '47. It was a single family house with three floors. Florence became pregnant again after the war. Florence said she had pains and I called the doctor. He told us to get to the hospital. I had our two little tikes here in the house and I couldn't leave them alone. I called a taxi and Florence went up to the hospital by herself. About three hours later I got a call from the doctor. He said, 'You have a little girl.' She was standard size. The twins were six when she was born. Florence called Shelia her immaculate conception, because I was never home. That was February 9, 1948. Shelia Mai, was named after my grandfather, Moishe. She was born in the Malden Hospital.

The third floor of our house was full of toys and couches, and an old Zenith TV was up there. I wanted the room cleaned out and after 20 minutes of carrying things downstairs, I opened the window and started throwing things out that I knew were junk. Out went the TV, the kid's old toys, an old couch, and we had that place cleaned out in no time. Florence and the kids thought I had gone nuts. We built a beautiful room up there. That's where Florence and I slept. The kids stayed in the rooms downstairs.

SUMNER REDSTONE was a young lawyer who had come from Washington to be with his father, Mickey. He owned a piece of a theater in Long Island near the Idlewilde Airport. He said our land in Revere would make a good spot for a theater. I called E. M. Loew and flew him around in my airplane. He said, 'No, the land is too low.' Sumner and I decided to move the dirt fill from across the street and made the property 'high land.' We built the first theater in 1948 in Revere. Loew became our competition. He had a theater on the Lynnway in Lynn. My partners knew how to go about connecting the speakers and setting up the parking lot. Mickey Redstone, Sumner's father, handled a lot of the details, and had some shares in the theater, too.

We gave out contracts and built the drive-in. We opened on August 25, 1948, premiering *The Swordsman* and *Her Husband's Affairs*. We had 800, or 1000 cars

OPPOSITE: JULIE AND FLORENCE DRESSED FOR A NIGHT ON THE TOWN. BELOW: THE GOLDMAN FAMILY. CLOCK-WISE: MYRNA, FLORENCE, JULIE, MARLENE AND SHELIA.

waiting in the street to get in all the time. First we had enough space to fit 1000 cars and then we made room for up to 1500 cars because we had so much business. The business was booming. It went like mad. We were open seven nights, showing two movies a night.

I noticed we were losing speakers left and right. After the war, speakers were hard to come by and people were stealing them. Back then, the speakers were hooked on to a cable on posts. The people put the speaker inside their cars to listen to the movie. When the movie was over they were leaving the speakers in the car. They'd drive off and the wire would snap. Factories couldn't make them fast enough. They were selling them to open air theaters all over the country. I thought we could lose every speaker we had.

I told Sumner Redstone that we ought to put in some steel aviation cable. I had a big roll of stainless steel cable used for controls in airplanes. The cable was the size of my finger and was very powerful. I bought micropresses from the telephone company. We cut ten foot lengths and drilled a hole in the cement. We put the cable through and micropressed it on. Then we drilled a hole in the speaker and micropressed the cable onto it. The kids couldn't cut through that with pliers. Then we put a sign up on the movie screen. It said, 'Please remove the speaker before driving away as you may damage your car.'

After the cable was attached to each speaker, we heard the sound of breaking glass after each movie was over. The kids didn't realize that they couldn't drive off now with the speakers. They broke some windows but we saved our speakers. We lost 170 speakers in a month and a half. We averaged 900 cars a night. Inside of a month we wouldn't have any speakers left. We stopped them from stealing. After that, RCA, who sold us the sound system, installed that same type of cable inside every speaker they sold. Kids couldn't steal them anymore and every theater owner was happy. Every post had a cable and everything we ran from post to post was underground.

Dick Berenson and I used to sell soda in the concession stand. We took plain water and poured flavored syrup into it. We sold it as fast as we could make it. We made orange and lemon drinks and sold it for 35 cents a drink. It was costing us about one cent to make. We built another big concession stand out back and made it very modern. We made an area for restrooms, baby warmers and baby changing stations. Families and teenagers came in droves.

One time I was parked in the front row with Florence and my kids. The movie ended but there was another show right after. I was watching all the cars drive out and I saw a car backing in. I said, 'What's he doing?' I jumped out of my car. In those days I was a tough guy. I banged on his window and said, 'What are you doing?' He said, 'Ah, we decided to see the second show.' I looked into the car. He had a bunch of guys in there. I walked to the back of the car. There were another two guys in the trunk. They hadn't paid to get in. They waited until the theater broke and then backed in to get in free. I opened the door and threw the driver out, over a six-foot fence. I grabbed another one and threw him over, too.

We built a golf driving range alongside of the drive-in theater. I was working the airport all day flying. When I finished I'd check out the golf range and then I would go into the theater and look around. We had a golf pro and his wife running the range. They had a big dog that was the size of a grown man. The dog would sit at the fence and keep people from coming in and stealing golf balls. The pro was going to go to a golfing banquet. He asked me to cover for him one evening. I was too tired to cover for him but he begged me. I said, 'Okay, I'll cover you tonight.'

We had two cash registers and thirty tees. I thought I better check the registers. I counted all the money in the register and I saw there was more money in there than what the register tapes called for. There was about $30 extra in one and $25 extra in the other register. I didn't think much of it. I saw Dick a couple of days later and mentioned that I found an overage of money in each register. He didn't say anything to me. About three weeks later the pro asked if I would cover him again.

I told him I couldn't cover for him all the time. I was tired from working so much. Once again I counted the register and there was $35 over in one register, $40 over in the other. This bothered me now so I mentioned it to his wife. We gave him the right to sell clubs and to take the best golf balls we had and sell them. She said, 'Oh that's my husband's money. He sold some clubs and just put the money in the register.' That made sense. Once again I told Dick Berenson and he didn't say anything. He just listened.

We had pinball machines at the golf range. We'd make as much as $10 a night from them. Golfers would

play pinball when they were waiting to get a tee. I came in one day and I saw this guy playing pinball. He put in $15 in the machine that day. I told my partner, 'We've got a guy who's throwing dollar after dollar into the machine at the golf range.' The machine was making us $30, $40 a night, and that was pretty good. About a month later, Dick Berenson said, 'Julie, will you bring the golf pro into town to the lawyer's office?' I said, 'Why?' He said, 'We want to talk to him about a new contract for next year.' I said okay and I arranged it. We went in there and they called him into the office. Dick said to me, 'You wait out here.' I said, 'Well, I want to listen.' Dick said, 'No, wait out here.' They went inside.

About five minutes later the pro came charging out screaming at me. 'You trapped me!' he said. They had a lie detector there, and they wanted him to take a test. He called me all kinds of names. Well, at first I was going to whack him. I was nuts in those days, and had a pretty good temper. I said, 'You go home by yourself,' and I took off in my car. Dick Berenson had liquor stores, and he had hired a shopping service to check on his tellers. The shoppers went into his stores, bought a load of booze, and they'd give the cashier a $50 bill. They'd watch to see how the cashier rang it in. Then they wrote about it in a book. The guy who was playing the pinball machine was a hired shopper. He would come in and give the pro $4 for so many buckets of balls and a couple of clubs. Then he'd notice that the pro would ring up $2. That was where he was making this extra cash. He was stealing about $300 a week.

There was no record of it. Only because I had counted the money had Dick realized something was wrong. The shopping service went to work. They told of how I was dressed and what I said to the customers, and they had a whole book on it. The pro didn't think I was smart enough to count the money. I only counted it to make sure that the other staff ringing up sales wouldn't take a buck or two. I didn't realize that he was the one stealing from us. Needless to say, we fired him.

THREE YEARS AFTER we opened the airport, a fellow came in and asked me if I owned the marsh out there. There were 100 or 200 acres of land. He said, 'I built a highway, C1, that goes out to Route 1 and I used some sand from that marsh out there.' I didn't know who he was. He was well dressed. He said, 'I'd like to have some more of your sand.' I said, 'How much will you pay?' He offered a nickel a yard. That didn't sound like enough so I said I'd let him know.

A week later another man came in. He said, 'Do you own that marsh out there?' He said, 'I'd like to get some sand from there. I'm bidding on a contract.' I said, 'Bidding on a contract for what?' He said, 'I can't tell you. Are you willing to sell the sand?' I said, 'I think I got it sold.' He said, 'To whom?' I said, 'I don't know who he is but he offered me a quarter.' He said, 'I'll pay you a quarter.' I said okay. I filled in his contract, and a month later he came in and said he won the bid.

The first guy was from Perini Corporation. They used to own the Boston Braves. The other one was

Martin DiMatteo Construction Company. DiMatteo won the contract, and he wanted to come in and start taking out the sand. I had to get a building permit. I couldn't just dig a hole in the marsh without one. I went to the building inspector. I said, 'I want to build a seaplane base off the end of my airport.' He said it was a good idea, and he gave me a permit without going to the councilman or anyone. Two weeks later, all these big draglines and machines came in. They stripped off the overburden from our land, which was quite thick. Now we were going to make a seaplane base.

There were 20 or 30 trucks in the morning and afternoon taking the sand to Boston Airport. Boston Airport didn't have pavement, it was just a little field. People burned coal in those days and they would dump the cinders out in the ocean on flats. When Boston Airport started, it was built on sludge from the harbor. Then they put about five feet of cinders over the top of the sludge. There were concrete slabs over sections of the cinders. When a pilot opened up the propeller wide there was a lot of suction from the prop tips. If you started the propeller the cinders would be sucked up into the prop and cause damage to the airplane.

There were eight concrete pads in a huge circle

around the airport. Planes positioned themselves to take off into the wind. We'd get on the concrete pad that faced the wind, open up the throttles and start taxiing down the concrete pad. Pilots would try to get in the air before reaching the cinders. In my airplanes I could always manage to do it, but the big heavy airplanes wouldn't be able to. By the time they reached the end of the concrete, they were rolling fast. The prop wouldn't throw the cinders as much.

To build Logan Airport, they had to build out in the ocean. They pumped up a lot of the ocean and they put my marsh sand on top of the fill there, and on top of

that they put gravel and then pavement. They made 10,000 foot runways out there. Sand was pumped up from the ocean and brought from my field. Gravel was brought in from New Hampshire. We got about $500,000 right away, for the sand. We paid off all the bills, and all the mortgages. We paid for all the airplanes we owed on and we still had money left over.

I WAS PRESIDENT of the Kiwanis in Revere. One of the Kiwanians said, 'We have an Italian world champ. Would you like to fly him around some places?' I said sure. So they paid me for flying Rocky Marciano. I didn't do it as hero worship. He was just a good guy to fly for. Rocky called me one day, I had just returned from Florida with Florence. We were at the airport with my airplane. Rocky said, 'Julie, I want you to fly out to Grossingers in New York. There's a girl out there I'm sponsoring as a singer.' I said, 'I have a cold. I just came back from Florida,' I didn't feel like I could make it. He said, 'Come on Julie, I need to get her to New York. I'm sponsoring her and I have a group of movie stars to hear her. Please fly us.' So Florence said, 'Let's go.' He said we could come along and stay as long as they stayed. I said okay. The singer was about 16 and the daughter of a Revere Kiwanian. He owned The Frolick Night Club in Revere. She came along with her mother, and her pregnant sister, a big blonde girl.

We took off in the airplane and by the time I got into the Berkshires area, it was rough. The pregnant girl seemed to be alright. I look back and the mother was sick. Florence had her head down in the back and she was trying to hide from the smell. I had the pregnant woman in the front. I couldn't get the wheel back, her belly was so big. She was sitting there and nothing bothered her. We got to Grossingers in the Catskills, and I parked the airplane on their little grass strip.

A few actors were going to be at Rocky's table, and he wanted us to go to the event. I said, 'I don't want to go.' He said, 'Oh, you'll have a few drinks.' I said, 'Well, I don't drink.' We got there and sat at a big round table with lots of stars. Eddie Fisher and Bob Mitchum were at Rocky's table. I had coffee or tea and the group was drinking up a storm. After they were drinking for a couple hours, Rocky said, 'Julie, do you want to pay for this?' I said, 'What are you talking about, Rocky?' I said, 'I didn't drink, I'm not going to pay the bill.' He said, 'Yeah, I'm not going to pay for it, either. Let's get out of here.' We left all the actors there and they got stuck with a huge bill.

We were pretty friendly together. After I flew him to a few different places he wanted me to come along with him. The trainers who used to spar with Rocky were famous fighters, too. He used to make personal appearances. He'd call me up and ask if I'd fly him to this city or that city. I always made my money for the flight. One day, we flew to Sanford, Maine, for a big party. We were picked up and put in two cars. I was in the back car and he traveled in the front car. There were three girls in each Roadster. We went to eat in the town, and they wouldn't take any money from us. I ate a steak. He ate two steaks. Then we went back to the hotel where

the mayor was having a party. They served lobster and chicken to everyone. Rocky said, 'Julie, I don't want to stay here too late. The mayor thinks I'm going to stay here all night. Figure a way to get us out of here.' I told the mayor, 'The weather is bad in Boston, and I've got to get him home. I wouldn't want to endanger Rocky.' The mayor said, 'I thought he was going to stay, but you better take him.' I got him home.

In September of 1955, Rocky was fighting Archie Moore in New York at Yankee Stadium. This was to be his last fight before retirement. I flew him down but then I flew back alone. Craig Smith, the vice-president of Gillette, asked me if I was going to see the fight. I said, 'No, I wouldn't ask Rocky for any tickets.' He said, 'I think you ought to go.' I said no again. About an hour later a messenger showed up with two ring-side tickets.

I had just sold my friend, Charlie Wolf, a twin-engine

airplane. I said, 'Charlie, you want to see the fight?' He said, 'Well, I don't know.' I said, 'We've got ringside.' He said, 'Yes.' We flew down and sat with Robert Mitchum, and another actor. Everybody was cheering and yelling. Rocky had a hell of a fight with this guy and he beat him by a knockout in the 9th round. I never asked Rocky for tickets, though. I wouldn't do it. He paid me and that was it. If he asked me to fly him there, I'd fly him and I'd go home. We worked together

ROCKY MARCIANO BECAME GOOD FRIENDS WITH JULIE AND FLORENCE.

until he retired as champ. I met his father and mother but I never kept in touch with them. Rocky went to Chicago for a personal appearance one day. A young kid picked him up in an airplane. They flew out of Chicago into a hill and they all got killed. Stupid.

CHARLIE PARKER was a friend I knew from before the war. He sold me an airplane when he worked for Intercity Aviation. I hired him as a salesman for Revere Airways. He was with me for a couple years and he never sold an airplane. In the process, he was friendly with people in Washington. He became the secretary in charge of the National Aviation Trades Association. He called me from Washington one day. He said, 'You know I owe you. You kept me all that time and I never sold a thing. There is an outfit building a new airplane in Florida.' He said, 'They just built a new airplane called the Jameson Jupiter and it's beautiful. Why don't you take on the distributorship for New England?' I said, 'Well, I don't know. It's a brand new airplane. I haven't seen it out yet. I'll think about it.'

Three months later he called me back. He said, 'Julie, you didn't make contact with Mr. Jameson and I told him you were going to call.' I said, 'OK, I'll go down and look at it.' Florence and I took the Bamboo Bomber and went to Florida. We had never been to Florida before together and it was beautiful. I found the airport in Deltona, 18 miles inland from Daytona. It was a little airport that had been built during the war. I landed and met the people and they wined and dined us. I looked at the airplane, and it looked pretty good. They said the test model was in Washington, and they didn't have a plane for me to take out but they'd arrange for that. Mr. Jameson asked, 'How many airplanes do you think you'll take?' I said, 'I want to learn a little more about it.'

We went home and two months later Charlie Parker called me again. He said, 'You never got back to Mr. Jameson. You're going to lose the distributorship for New England.' I said, 'Well, I never got to test it out.' Charlie said a famous test pilot was flying it around in Washington, showing it to people. I flew down to Washington and I met the test pilot. I met him and he told me to get in. The wind was blowing like mad. I took off and it got in the air quick. I flew around and it handled all right. It was on tricycle gear, and had three seats side by side. It had an electrical system and radios so I could fly at night. I landed it, made another takeoff and landing and went home.

I was home a couple of months when Charlie called me up again. He said, 'Julie, you're going to lose the distributorship for New England.' I said, 'Well, I don't know what to do.' He said, 'Why don't you just try it.' I flew down to Florida and arranged the deal with Mr. Jameson. They were manufacturing the planes and things were humming in the factory. I was still worried so I told him I'd give him cash for two airplanes, and put the money in his bank under escrow for the other ten planes. I put it in a bank in escrow down in Deltona, enough for twelve airplanes so if they weren't delivered, I could get my money back from the bank. Everybody else gave

Jameson their money upfront, and he was selling distributorships all over the country.

When I got back, a fellow from Vermont called me. Earl Maxim was a farmer who had a small strip and a mink ranch. He said, 'Are you the fellow who's going to take on the distributorship for New England, for the Jameson Jupiter?' I said, 'Yes.' He said, 'That's strange because I got the distributorship.' I said, 'I'd like to meet you.' He was in Boston buying whale meat for his minks. He came right over to my airport and introduced himself. He was a tall guy with a Yankee drawl. He said, 'I got the distributorship.' I said, 'I don't think you have because I already made arrangements for the distributorship.' He said, 'Well, don't you think we ought to call him?' I said, 'Sure.' I had a big desk and he sat across from me. He listened on a second phone while I called Jameson himself.

Mr. Jameson said hello and told me that things were going great. I said, 'How many distributorships have you sold in New England?' There was silence. He said, 'None, except to you.' I said, 'Well, that's strange. There's a man sitting in front of me who says he has it.' He said, 'Oh that fellow, he lived down here in a trailer.' Maxim stayed in Florida in the winter time there. Jameson said Maxim had come in and asked about the distributorship but he never sold it to him. He had sold it to me. Jameson said, 'This guy only had a mink farm and one little strip in Vermont where you have an entire airport.' I looked over at Maxim. I covered the phone and said, 'Do you hear what he's saying?' Maxim said, 'Liar.' I said, 'Well, we've got to do something.' I talked to Jameson again. I

JULIE AND
FLORENCE WITH
MR. JAMESON.
THE JAMESON
JUPITER WAS A
NICE PLANE BUT
IT NEVER GOT
LICENSED.

said, 'Are you sure now, Mr. Jameson, that nobody else has it?' He said, 'That's right.'

Evidently, Jameson sold the farmer from Vermont a distributorship as well. Maxim had probably paid for half a dozen airplanes but he had no way of getting his money back. I said, 'This is crazy. I'm going down there.' Maxim said, 'Can I go with you?' I said, 'If you pay your way.' He said okay. I had a client who had a trucking business and wanted to get a multi-engine rating. I asked him if he'd want to go on the trip for multi-engine time.

The next morning all three of us climbed into my airplane and landed down in Daytona. I went to a lawyer in the town and told him the story. I said, 'I don't want you to sue for anything, I just want some advice and I'll pay you off.' He said, 'Okay, I'll bill you for whatever my time is.' I told him the story. He said, 'You have a good case against him if he sold that franchise to somebody else before he sold it to you.' I told him I had my money in escrow. He said, 'Well, you can get your money back.'

I had infected my nose and it was swollen. That was all I could see. I went to the nearest hospital. The doctor saw me and said, 'You can't go anywhere, you've got to stay in the hospital.' I said, 'No, I can't.' We were getting a hotel room when Maxim invited us to stay at his trailer. It was hot in there, it was in the spring or summer. I laid down in there and I thought I was going to die. My nose was throbbing.

We rented a car and drove in the front yard of Jameson's operation. We didn't tell him we were coming. I said, 'If anybody goes out, stop them.' Earl Maxim went in the back door, the trucker went in the side door, and I went in the front door. It was a big hangar but I heard someone inside once I went in. A girl came up and said, 'Can I help you?' I said, 'Yes, I'd like to see Mr. Jameson.' She said, 'He's not here.' I said 'Well, I'm going in the hangar.' I went to a smaller door and into the hangar. People were working out there, and the big doors were open. I saw Jameson heading for the big doors. In walked my man from Vermont and Jameson stopped. He turned around and saw me again and said, 'Can we go in my office?'

There was just the four of us. I told him, 'You sold me a franchise and him a franchise. You took my money and you took his money.' He started to cry. He said, 'Well my salesman got me into trouble. He went out and sold all the franchises and he shouldn't of done it.' I said, 'Well, that doesn't make any difference. You sold the same franchise to two of us.' He said 'Oh, it's all yours, this guy just has a little strip and...' Maxim interrupted. 'What do you think I'm going to do?' he said. I said to Jameson,

'Look, I want my money back. He can have the franchise.' Maxim said, 'I want my money back, too.' There were tears coming out of Jameson's eyes. He said, 'We are selling a franchise on the other side of Florida. We'll have that money here in a day or two.' My nose was big. All I could see was Jameson and my nose. He said to both of us, 'I'll give you your money back as soon as I get that.'

We left and I went to the bank I had an escrow account in. I told them not to release any money unless I approved it. I went back to the trailer at Daytona and laid there for two or three days. Jameson came to the trailer and said, 'I got your money for you.' It was a regular check. I made him get me a bank check. Then I said, 'For all my flying and the bothering with you, I need some money.' 'He said, 'How much do you want?' I said, 'I want $1500 to cover my expenses.' He started to cry again. He said, 'I'll give you a check.' I said, 'Well you better not let this check bounce.'

We flew off to home and I never cashed the check. I got all my money back and I learned my lesson. The Jameson Jupiter never got on the market. They built about 50 of the planes but they never got licensed. It needed a certain amount of retesting and it was underfinanced, so it never worked. If he had the money he would've had no trouble at all. Some other people came to me to invest in a seaplane. Everybody told me to buy into it. I didn't do it and it never got licensed either. I was approached two or three times but I said no thanks.

SAMUEL C. JOHNSON of Johnson Wax was a student of mine. His grandfather called me from their headquarters in Racine, Wisconsin. He said his grandson was coming in with an airplane and he wanted me to teach him how to fly it on the water. He had been checked out on land already. The plane was a Grumman Widgeon, an amphibian for both land and sea. The Widgeon wheels came down on the sides and out of the nose. Hydraulic legs pushed the wheels out to land on runways and pulled them up for landing on the water. Pontoons were at the end of each wing, so if you landed and a gust tipped the wing into the water it would float.

We went out to the harbor near Salem and Beverly and landed in the water. I had Johnson dock on the buoy and he did alright. We smelled smoke. I looked back and smoke was coming out of the floor. He said, 'I had some radios put in before I came. Maybe the wires have short-circuited.' The smoke was coming from the hull.

I didn't know how to swim so I decided we'd go to the coast guard base in Salem. I opened the throttles and we rushed toward the base. The airplane became buoyant. We were in the air, so I thought it would be smarter to come straight back to my airport. I landed on my runway. We had the seaplane base but that was too big of an airplane to land it there. Our seaplane base was too short. The smoke started when the radio wires had chafed and started to burn. We fixed the problem and I gave him more training in the water.

Johnson went to Harvard, graduated and left Boston eventually. His grandfather and father died and Sam became the CEO of Johnson Wax. I saw an article about him in a magazine. On the wall behind him was a picture of the Widgeon that he learned to fly in. He owns a three-engine French jet now, called a Falcon.

IT WAS SNOWING like hell. I was in the office when I heard the roar of engines. There were two P-51's circling over the airport. Both pilots had their hatches open, and they were 800 feet above me. I thought they must be lost. I got a board and drew B O S in the snow and made an arrow pointing towards Boston. I jumped in my fire wagon and hoped they were good pilots. I was sure they couldn't stop on my little runways in the snow. They landed and were sitting at the ends of two runways.

I climbed up on the wing of one plane and said, 'What are you doing here? This is Revere.' They thought they were at Westover Air Force Base. They were on their way from Indianapolis to Bangor, Maine, to train in a

Scorpion jet fighter. They had to land being low on fuel. They couldn't see Logan Airport because of the snow. I didn't have the type of fuel they needed. I took care of them and put them up in a hotel. They waited until the weather cleared, then went to Logan for gas. From Bangor they sent me two pen sets as a thank you for my help. They were very capable pilots.

A C-47 LANDED at our airport one evening when I was plowing. Revere was the only airport open due to the snow. The crew chief came out of the airplane. I said, 'By the way, it costs $5 to land here.' He argued with me and said, 'The Secretary of the Navy is on board. We pay nobody.' I said, 'You've got to pay five bucks.' The pilot came out next. It was Frank Schwikert. He was one of the pilots in my outfit during the war. He said, 'What the heck are you doing charging me money? It's a Navy ship.' I said, 'Well, so long as it's you, I won't charge you anything.'

Frank Schwikert was in the Wilmington outfit with me. He became a pilot for General MacArthur before and during Korea. He used to fly through Newfoundland all the time. He was also the personal pilot of the Secretary of the Army, Kenneth Royal, after the war.

They got out of the airplane. Cars came and picked them up. It was a C-47, just like the one I flew around the world. That Secretary was the same man who got shot when Jack Kennedy was assassinated in Dallas. He was the governor of Texas at that time, John B. Connally, Jr. He was wounded badly but at least *he* survived.

A GOODYEAR BLIMP parked in our airport one day, advertising Wonder Bread. Somebody shot the damn thing full of holes from a nearby hill. The Goodyear people saw the pressure going down and they patched it.

Another day, after that incident, I went out with a student. We were flying over Revere and I had a thought that something was wrong but I didn't know what. I came back and landed. They were checking the airplane when they saw the hole in the bottom. Somebody had shot a bullet into the airplane. It had gone behind my back into a round tube in the seat, and buried itself in the tube.

We never found out who shot the gun. It had to be someone around that area. If the bullet hit one inch further the other way it would have gone into my spine.

ANNE BRIDGE WAS A woman who came in to learn how to fly. I had an instructor, Whitney Welsh, who was giving her lessons. She was making some landings and bouncing a lot in the Taildragger she was flying. Planes with tricycle gear have a nose wheel but Taildraggers have a tail wheel. I told her to get the tail down but she said it was hard for her. She was a small girl. I got two or three heavy bags of sand and put them in the baggage compartment. When she pulled back on the yoke, the weight balanced and she was able to make good landings from then on. Anne entered races in the women's aviation club. She raced across the country and won awards.

Through the years she met Dr. Baddour, a professor at MIT. He was involved in the development of Ipogen,

intravenous iron and nutritional supplements. Dr. Baddour worked with MIT and the Draper Lab. Dr. Charles Draper started his work with gyroscopes, making guns more accurate for World War Two. Then he invented the inertial guidance system. It takes a plane all around the world, no pilot needed. The system is used in submarines and space vehicles now, too. He was quite famous and considered the father of inertial navigation. In the fifties, my sister's husband, Sid Pliskin, worked on these guidance systems with Dr. Draper at M.I.T.. The Draper Lab still exists today in Cambridge.

JOHNNIE GRIFFIN was going to a Quiet Birdmen meeting, and I asked if I could go. I had heard it was a special club and wanted to learn more about it. I attended several meetings and then was recommended for membership in 1950. Eventually I became the beam man. For every meeting I wrote up a report and sent it to the QB's magazine. I was later recommended to be the key man at Boston Hangar and I did that for a year.

I made several friends in the QB's. Members were famous pilots and chief pilots. They were the best in the region. Johnnie Griffin started East Coast Aerotech. Jack Phillips started out as a line boy at Muller Field. He gave me my aerobatic flight training. He used to drive to Maine with me in the trucking business. Lee Hipson gave me my aerobatics flight test. He was an examiner for the CAA. Chip Collins was a pilot during the war. He piloted the first flight test for the inertial guidance system that Draper Labs created. He and Dr. Draper flew to

California and back with the inertial guidance system.

Chet Matika was a friend who owned a pilot outfit in Plymouth Airport. He is on the QB national committee. Ken MacDonald flew with us in Revere. He took lessons with us and I gave him his flight test. Then he became a pilot in Bedford, at Hanscom Air Force Base. Joe Garside was the head of Wiggins before the war. Frank Cummeford was an instrument examiner, one of the best in the country, at Hanscom Field. Link Noble flew at the Mansfield Airport. Henry DiMassalle flew for me as an instructor and examiner.

WYLIE APTE owned the White Mountain Airport in North Conway. He was a good friend. I met Wylie through the Quiet Birdman. He was called the 'Old Man of the Mountain.' One day Florence and I took the twins and drove way up to northern New Hampshire in my Chrysler. I blew a tire north of Conway and drove back to Wylie's place. I couldn't find another tire of the right size anywhere. My car had big fat tires. I borrowed a plane and flew to Boston for the spare, while Florence and the kids stayed up there.

Wylie had a bad fire in his hangar one time and almost lost everything. I donated money and a prop and other things to get it started again. Wylie was a nice guy. He never charged me a dime for landing, although there was always a landing fee there. He opened it in the 20's and it stayed open until after his death in 1970. His son, Bunkie, ran the place after he died. Bunkie was a pilot for TWA. He sold the airport property and has a big home on one of the mountains there. Across the sky east of Conway is the Wylie Intersection, named by the FAA and used by pilots tracking in on the federal airways.

AL LECHSHIED went to work as a pilot for *The Boston Post* after the war. *The Boston Post* owned a Lockheed Lodestar, a big twin-engine airplane, and it was a taildragger. Al became the chief pilot. Al's copilot was Ken McKensie who had also been an instructor for me. One of my instructors, Whitney Welsh, used to fly with them in order to get training in a big airplane.

Lechshied and McKensie did a lot of trips to Newfoundland. They went up to the northern part of Canada where there were cobalt and chemical mines. John Fox, the owner of *The Boston Post* and part-owner of Western Union, invested in the mines. Al had invested all of his salary and everything he had into the mines. He wanted me to invest in it too. In order to enthuse us about buying it, the boss told them to fly me up to where the operation was near Hudson Bay. All the guys at the airport wanted to go. The plane fit 18 passengers so they took 18 guys.

We landed up on this little strip and they took us to the mines which went way down into the earth. They

ABOVE:
FLORENCE AND
WYLIE APTE.
THE AIRPORT
HAD ITS OWN
'OLD MAN OF
THE MOUNTAIN.'
OPPOSITE:
CHARLOTTE
KELLY AND ANNE
BRIDGE IN
FRONT OF THEIR
CESSNA 120
TAILDRAGGER.

were blasting cobalt, silver and nickel. We went down the shafts and they explained what they were doing. There was a smelting plant where they'd brush off the stone and take out nickel and cobalt and silver. We saw that and we all came home and everybody bought like mad. I was very conservative but I bought 2000 shares. Everybody at the airport heard I bought it, so they all

OPPOSITE: AL
LECHSHIED,
JULIE AND KEN
MCKENSIE IN
FRONT OF JOHN
FOX'S LODESTAR.

bought shares too. If they had $100 they bought cobalt and chemicals.

I bought my shares for $1.10 a share and sold them for a nickel. At that time they didn't need cobalt that much and they were getting a lot of it from Africa. Middle Africa had much more than anywhere else in the world. The supply was so much that the price wouldn't go up. The silver prices didn't go up until years later. When they started to build jet engines all around the world, cobalt went out of sight. They were making jet blades and they ran cherry red. The only metal that runs cherry red and still maintains its structure is cobalt.

I was flying my Bamboo Bomber and doing charters. I had just flown a charter into Northern Maine when I got a call there. It was John Fox with Lechshied and McKensie. 'What are you doing there? We need you,' Al said. They were heading for Newfoundland and wanted me to go with them. I had a charter with people who were bidding on a big project in Canada. He said, 'Come home, come home.' I didn't go home. I stayed with the people and with the weather beating bad I got stuck for a day. I finally got back into Bangor and cleared customs. John Fox's group was there without him. They had the Lodestar, and a man from *Life* Magazine was aboard. I sent my airplane back with one of my instructors and I went on to Newfoundland with them.

I knew a lot about Newfoundland because I had flown over it for 14 solid months. McKensie and Lechshied were going to talk to a man named Noseworthy from Canada. He had drilled a hole for a water well and he brought up oil. It got back to John Fox. Fox decided he was going to try to get oil up there. He made all the arrangements and brought bulldozers and drilling equipment to a site. We left Al up there. McKensie and I flew Noseworthy to Montreal and then we flew home to Boston. Al came back from Newfoundland eventually. John Fox never found oil so he quit trying.

Fox had an estate in Maine, with a grass strip on it. Fox told Al that they were going to fly up to the strip at noontime on a Saturday. Al said okay and went to the airport to meet Ken McKensie. John Fox never cared about anybody else. He always did what he wanted, when he wanted. You could wait two weeks. Al and Ken waited five hours for Fox and the weather had gone bad. They didn't have facilities on the Maine strip for bad weather. Al and McKensie went home. Al got a call an hour later. John Fox said, 'Where are you? We've got to head for the strip.' Al said, 'We waited all day. You said you'd be there at noon and we left at five.' Fox said, 'I put all those radios in the airplane so you could navigate with it in bad weather.' Al argued that he had no facilities to navigate into, and the weather had to be decent enough to see the land. Al said, 'I'm not going to go.' Fox said, 'You will go.' So Al said, 'I quit.'

Fox called up McKensie and wanted him to be the captain. McKensie didn't have Al's experience but he was a good pilot. They didn't go to Maine that day because they needed a copilot for McKensie. They asked one of my instructors, Whitney Welsh, to become the copilot. Whitney Welsh had no multi-engine time. He shouldn't

have been there as a copilot unless he went to a multi-engine training school. Fox wanted to go to Philadelphia and he was having money troubles. He invested and things didn't go. He didn't pay for his gas at Logan and they cut him off. He had enough gas to go to Pennsylvania. They dropped off Fox and they were to go back home. McKensie knew that their credit was no good at Boston. He landed in LaGuardia and they filled the airplane up with gas, no questions asked. A captain, copilot and an engineer from Northeast Airlines had missed the last flight going to Boston. They asked McKensie for a ride and he said sure. He had no one aboard except Whitney Welsh. Captain John Mudge from Northeast sat in the copilot seat with McKensie as pilot. Mudge's copilot sat in the back with Welsh and the engineer.

My mechanics had put the engines into Fox's plane. The first engine we put in had water in it. My chief mechanic said we had to take off all the cylinders and clean it out. Fox didn't want him to do it, but he did it anyhow. He found the problem and he fixed it. Fox complained bitterly about it because of the bill. They flew the engine about 300 hours and it seemed to be alright. We put in a second engine that came from the same place, off a dock in East Boston. Bob Adair who bought junk had bought the engines and sold them to Fox as good engines. When they were originally put on the dock, they might have been all right. But they could have been wet with sea water or salt may have gotten in there. We wanted to take the second engine apart and Fox wouldn't let us.

When they got over Connecticut, McKensie and Mudge went into the back and Welsh and the copilot from Northeast went in the front. Whitney was in the pilot's seat when the left engine blew. Whitney didn't have the knowledge of how to keep the airplane straight when that happened. The plane started descending in a big turn. They weren't really high, only about 3000 feet. McKensie ran and switched seats with Whitney. The three guys were in the back, McKensie was in the left seat, and Whitney got in the right seat. They descended and they couldn't hold it. By the time McKensie had taken the seat it was too late. They went into a big hayfield and hit some trees. The plane caught on fire in the crash. It split in the back. Mudge punched a window out in the tail and his two guys got out. A farmer came out and asked Mudge if anybody was still in there. Mudge said the two pilots in the cockpit were there. The farmer ran and pulled McKensie out. Whitney Welsh was dead. McKensie was in the hospital for many months, burned all over. He eventually died from it.

I was given a subpoena to an FAA meeting at a Northeast hangar. The FAA wanted to know what everyone thought about the accident. I said, 'You should never put a used engine that hasn't been overhauled in an airplane. Especially when there are no records of it. The engine was sitting on a dock. Even if it was overhauled, it should have been opened up and redone.' Everybody got mad at me because they were all buying engines from the military. It wasn't cheap to rebuild an engine. But records were needed for each engine. Al

Lechshied worked for a long time in the aviation business and then he went to work for me.

I GAVE MY PARTNER, Dick Berenson, some flight training when we were in the State Guard. After we opened the airport, he wasn't interested in learning to fly. I told him I wanted him to learn how and he said, 'I don't want to.' I said, 'Why?' He said, 'I've got you to fly me around.' I was flying him around everywhere. He knew some Jewish people who owned the Mount Washington Hotel. The owners knew a beautiful dame. They wanted him to meet her. He was single. I called the state police and asked if I could land up there. They didn't advise it. But he wanted to go so I went anyhow. I flew him up to Mt. Washington to a little strip. It wasn't really big enough for my airplane. I got in there but it was tight.

We went into the hotel and I had never had such luxury in my life. They gave me a private room and I had lobster and steak. The woman they brought to see him was a gorgeous creature. Entertainment was going on everywhere. Downstairs they had unlicensed gambling halls and slot machines. Everything was going full blast when we got there.

The owners treated me good because they knew I was Dick's partner. We stayed overnight. I don't know how Dick and the dame made out but he never married her. We flew out of there the next day. I had a hard time getting out. The runway was too short. I ran from one end of the field to the other and got it off the runway by the skin of my teeth.

Once in a while I flew Dick to the Cape. He had a home in Falmouth, Massachusetts. I took Florence and the kids down there two or three times. Louie Fox flew with me a few times. Fox was called a sportsman, which meant he was strong in the city, in underground stuff. The police never had anything they could pin on him though. Mickey Redstone flew with me when we were looking to start the open air drive-in theater. I let Louie Fox fly the airplane a little bit and Mickey Redstone went wild. He yelled, 'Get down, you're going to kill us!' I had to come back and land. He was so nervous with Louie flying but I had the dual controls on. Louie couldn't do anything I didn't want him to. Doc Sagansky never flew with me. He didn't fly in small airplanes.

DR. SAGANSKY was a dentist and a head bookie. He used to back a lot of bets. The betting pool was based on the numbers of the treasury. It started in the New York ghetto. There were a lot of bookies and one boss. The treasury is printed in the paper every day. The last three numbers were significant for the pool.

I never bet on it in my life, but a guy came up to me when we had the trucking terminal in Boston. He said, 'You want to take a ride on the book?' I said, 'No I don't gamble.' He said, 'It's a nickel.' I wouldn't do it. He said, 'I'll put it on for you.' I won, so he gave me $25. After that I wouldn't bet another nickel.

Doc backed all the bookies. If a pool writer had a lot of people win, he'd go broke. If a bookie wrote $200 in bets, the bookie would pay Doc a percentage, like 10% of total bets, or $20. A bookie might write $1000 worth and then give $100 to Doc. If the gamblers won the bet, Doc would cover the bookie, but nine times out of ten they lost.

One day I called Doc. I wanted to bet on a certain horse I knew was going to win. He wouldn't take the bet. He said, 'Julie, chances are you are going to win. Then you're going to be spoiled. You're going to be spending your money on backing horses.' He said, 'It's no good, stay out of it.' He profited because he was on the top tier. He was a good guy, he liked me. He trusted me.

When I walked out of his house with $50,000 in my pudge, I said, 'Where do I sign?' He said, 'You don't have to sign.' I said, 'Why? I couldn't get $500 without two cosigners, when I wanted to get an aerobatic rating.' He said, 'I talked to you long enough. I know you're okay. If you weren't okay, you could sign a million times, it would mean nothing. I wouldn't give it to you. You're okay.' He trusted me so I trusted him.

I GOT A CALL ONE DAY from a woman who had a butchery up in Haverhill, Massachusetts. She said, 'Are you Goldman from the airport?' I said yes. She said, 'Well, there's something strange going on. Two little girls I had visiting said that you were their uncle. Their last name is Kyle. I brought them over from the Christian orphanage before Christmas. I fed them and did all kinds of things with them. These two little girls knew all about challah, Jewish bread. They have to be Jewish. What are they doing there?' I said, 'I'm not sure. I don't know anyone named Kyle.'

Uncle Iz came in. I told him the story. I said, 'You think it'd be Saul's kids?' He said no. I said, 'Let's get in the car.' We drove to Haverhill and got to the orphanage. We came to a huge brick building. I knocked and a nun came to the door. I told her who I was looking for. She said, 'We can't let you in here.' I said, 'Well, can I talk to your boss?' She said, 'Oh, Mother Superior. You wait here.' When she left I pushed the door open. There was a huge hall and the nun was running down toward the other end.

The nun and the Mother spoke in the hall. She said that I had come for those kids. The Mother came and talked to me. After talking, she brought the children to the door. Sure enough, they were Saul's girls. They grabbed me around the legs. They said, 'Uncle Julie, Uncle Julie.' I said, 'Can I take them home with me tonight?' She said, 'Oh no, they have to say their catechisms in the morning.' I said, 'I've got a partner at the airport and he's a strict Catholic. He'll take care of them for Sunday morning. I'll do anything you ask me to do. I just want to bring them home.' She let me take them home.

I got into Malden, and I was boiling mad by then. I called Saul in California. I said, 'What the hell did you do? You put your kids in an orphanage?' I said, 'You were in an orphanage and you got sick, you almost died there.' He said, 'What are you talking about?' I said, 'Your kids are with me right now. I took them out of that orphanage in Haverhill.' He said, 'Wait a minute.' He talked to his wife, Bea. Saul had a sub shop in Watts, and another one in Long Beach which he opened for his son. For a long time he wasn't doing good at all. He and Bea started to gamble and they gambled away everything they had. It was terrible. They sent the girls to their grandmother, Ms. Van Buren in Maine. The boy went to his aunt's house in Illinois. Bea left the little boy with the aunt but took the two girls to this orphanage. She never told Saul. Evidently, the grandmother in Maine said no. Saul said, 'I'll put Bea on a train. She'll be there in three days.' Bea arrived three days later and took the kids and they went off. I didn't think he should put his kids in an orphanage. I knew what that felt like.

Saul had changed his name to Kyle. Bea was a French girl who he met in Maine. When they married and moved to California, she didn't want anyone to know they were Jewish. They took part of Gukailo, our family name from Poland, and changed their name to Kyle. He went from Saul Goldman to Mike Kyle.

Years later, Florence and I went to see Saul and his wife. He drove us down from Los Angeles to Ramona where they had a second place. All of Saul's kids were there. He had three girls and a boy by then. They wanted a ride in an airplane. A little airport was near there. It had a load of Bamboo Bombers parked on the field. They were used to dump boron on fires. I said I'd go down and check if they have an airplane to rent. They said, 'We'll rent you an airplane but we'll have to check you out in it first. I said, 'I don't mind.' I gave them my license with my name on it.

I didn't think anything of it. I got the airplane and flew it for the check ride. Then I took all the kids for a ride. When I came back down, Saul said, 'What'd you do to me?' He said, 'You told them that you were Jewish?' I said, 'I didn't tell anybody anything. I gave them my license.' I couldn't get an airplane without my license. He said to me, 'Well, I'm in trouble.' Ramona, California was a very, stiff town against Jews. They never let anyone know they were Jewish and he said I ruined it for him. It got around that he was Jewish. His wife disliked Florence and I so much for that. I didn't care. I was born a Jew, I'm going to die a Jew. I'm no different than any other guy in this town.

Many Jewish pilots in New England changed their names. They couldn't find work with an airline unless they had a name like Lyons or Lynch. Any name would do but a Jewish name. There were about seven or eight Jews working for Eastern who had changed their names. The only names that were on the door were McCarthy and Ryan, and every other one was a Jew. Dick Berenson wanted me to change my name but I never did.

CROCKER SNOW HAD A Navion, a tricycle gear plane. He used to fly it out of his strip in Ipswich. As he came down toward Logan Airport, it was all fogged in. They had too many big airplanes landing, and couldn't clear him in to land. Crocker was the Director of Aeronautics for the State and a very capable pilot. He headed up the highway to our place and landed, I didn't see him land. That's how bad the weather was. He taxied up in front of my building.

Just then, I got a telephone call from the tower at Logan. A man said, 'Did you have an airplane land there just now?' I said no. He said, 'Well, something came down there.' I said, 'I don't know what it is. I'll go out and look.' I looked out the window and saw a shadow. A dark blue Navion sat there like a ghost. I knew it was Crocker Snow. I went back to the phone. They asked again if I saw a plane. I said, 'Well, what about it?' The guy said, 'Well, he broke all the rules. We want you to file a complaint.' I thought, 'Not on Crocker.' I said, 'Why don't you come down here and write the complaint. I won't complain, the hell with you.' I wouldn't do it.

Years later, we both got an award at the Aero Club. I told the story of how he landed zero zero on my little field. He couldn't see forward, and he couldn't see down. He still flew in it, he was a very capable pilot.

I WAS PLOWING THE runway after a snowstorm when a fellow walked in. He wanted an airplane ride and paid me $3. My brother-in-law, Victor, was plowing as well. I told him to make another cut on the runway to make it wider. The man went off with my pilot, Ken, and we waited for the plane to come back in. The plane took off, went around and flew by us. I said, 'Someone is standing in the back of the plane. That's the guy who bought the ride. What is he trying to do?' It looked like Ken was trying to control the airplane. I could see this guy in the back standing up. Ken was all over the place. All of a sudden, something fell out of the airplane.

It went like a torpedo down into the marsh. Ken flew by and landed. I drove onto the runway. The runway was higher than the marsh about seven feet. I went to the marsh and looked down. There was the fellow lying on his back.

As I looked down his eyes opened. I said, 'He's not dead!' Roy Swain was a new instructor who worked for me. We jumped into the marsh and rolled him up onto the runway. I said, 'Don't bend him, he must be all broken up.' I had a banana pad in the trunk. I got it out and told Roy to take my car and go call an ambulance. I said, 'He'll die down here.' I got him on the pad and laid him on his back. Then the ambulance came and they took him away.

I asked Ken, 'What were you doing with him?' He said, 'I was grabbing his foot. I thought I'd hold him and land.' I said, 'You'd kill him on the landing.' He said, 'No, I was going to fly over the seaplane base and drop him in it. But I couldn't hold him at all and he jumped!' When they talked to the passenger, he said that he fell out of the plane. The way this airplane was built, he couldn't have fallen out. I said, 'He jumped out.'

About six months later in the paper, the headline read, 'Jumping John Does It Again.' The same guy went on a Nantasket boat across the bay. He climbed up the mast and jumped into the ocean. One of the chief sailors jumped in after him. He was trying to save him and Jumping John bashed him. He tried to drown the sailor but others pulled them both out of the water. At the time he was suing me for $50,000. After that happened his lawyer never contacted me again. He was a nut.

I TAUGHT CRAIG SMITH of Gillette how to fly. I was told by Gillette's chief pilot to teach Craig how to fly in a Tri-pacer. The chief pilot said he wanted to solo Craig when it was time. I gave Craig 10 hours of training and was ready to let him solo. I said, 'I'm going to let you put in a few hours of solo. When your guy shows up, I'll tell him you're ready to solo. But don't tell him I did that.' Craig soloed for a few hours before the chief pilot showed up. The chief pilot sent him on what he thought was his first solo flight. He said, 'By God, he did a good job.'

Craig wanted to do an ad with aerobatics for a commercial during the 1953 World Series. The commercial's slogan was, 'You can take a shave as quick as you can do a loop.' Craig had called Piper and told them Gillette wanted the commercial done there in exchange for a free plane. Gillette had sent a pilot to the Piper Cub factory in Lock Haven, Pennsylvania, to film the loops. Whatever happened, the pictures didn't work out. Craig then asked me to do it. I didn't want to use their tri-pacer airplane. The tri-pacer was not an aerobatic airplane. But I said, 'If they can do it, I can.' The photographer sat in the right seat and I sat in the left. I did four loops, one in each direction, and was paid $400. The pictures came out. They aired the commercial during the World Series. I recognized my gloved hand in the film. That's all I could see of me.

Craig and I became good friends. We traveled to aviation events together. There was an aviation meeting out in Phoenix. I said, 'I'll get a room when I get out

there.' He said, 'You'll find you won't.' I went out there without him and I couldn't get a room. I called one of the big hotels that I thought Gillette would use. I asked, 'Do you have a reservation for a Mr. Goldman there?' They said yes. It was a nice suite compliments of Gillette.

I GOT A LETTER from the president of Boston & Maine Railroad one day. It was addressed to the president of Revere Airways. Inside was a picture of our airport from overhead. It was a picture that I recognized. It had been photographed from my plane. The letter said it was possible to build a railroad siding on the airport property. We had a railroad track at the end of the marsh.

JULIE, REPRESENTING REVERE AIRWAYS, RECEIVES AN AWARD FOR SAFETY FROM CROCKER SNOW AT THE AERO CLUB.

The letter detailed how many railroad cars could come in and out with freight per day. The railroad didn't realize that the man who inquired with the picture was Dick Berenson, the treasurer. He didn't want to be president. He wanted to handle the money.

I didn't say much to anybody, and one day Dick and his father came in. Dick said, 'Julie, I think that the aviation business is slowing down. I think we ought to sell the place.' I said, 'What do you mean?' He said, 'Well, we can get a good buck for the place.' I said, 'Well, how much do you think?' He thought we could get $200,000. He owned half. I said, 'You want to sell for $200,000. You'd be satisfied with that?' He said, 'Yeah, and we have a lot of money in the bank.'

We had all the airplanes paid off. We had no mortgages—that seaplane base paid for everything. This wouldn't include the planes or the equipment, that would be separate. I said, 'Are you sure you want to sell it for $200,000?' He said yes. His father was standing beside him. I said, 'I don't think we should sell. We still got a lot of years left in the business.' He said, 'Julie, I'm telling you we're losing money now. The flying has slowed down.' I said, 'Are you sure?' He said yes. I looked him in the eye and said, 'I'll buy you out.' His father turned red and left the room. Dick looked at me for a while. He didn't say another word, and he left. He was stunned.

I called up Doc, my backer and friend from the beginning. I said, 'Doc, I did something, and I hope I get your backing.' He said, 'What is it?' I told him the story and said I offered to buy Dick out. I said, 'I'll

need at least $100,000 in cash. He said, 'Whatever you do is okay, Julie.' Dick never, ever, spoke a word to me about selling out after that.

By 1960, the partners had shares in three theaters. I was a partner in the Suffolk Drive-In Theater. Then I invested in a third theater in Avon, with the Mount Washington Hotel owners and the Berensons. Before I knew it, Sumner Redstone was running about 40 or 50 theaters of his own.

The Revere Drive-In became Showcase Cinemas, on the same spot in Revere. Sumner never took in the same partner for any two theaters. He was a pretty clever guy. Nobody could control his operations except him. Sumner called his company National Amusements, Inc. Today he owns Viacom and Blockbuster and is one of the richest and most savvy businessmen in America.

The State took the golf course land by eminent domain for part of the highway, Route 1. They also took part of the airport. I worked harder at the airport than I ever would have worked for the airlines or any business. Commercial pilots flew 85 hours a month. I was flying 120-130 hours a month or more, in addition to managing all the administration of that little airport. I had a good reputation.

Dick had a proposition to build a shopping center on the airport land. The Berensons had shopping centers all over the country, called Dreyfus Properties. I bought everything from the airport, all the planes and equipment down to the last bolt. We flew 139,000 hours from the time we opened in 1946 until we closed in 1961. Then I

moved the operation to Beverly Municipal Airport. I changed its name to Revere Aviation.

Northgate Shopping Center was built in 1961 on the old airport land. I became a partner. We had a terrible time keeping the shopping center alive. Without my knowledge, the partnership administrators signed some bad leases that almost broke us. Some of the leases ran too long with too low of a rental. One lease was signed for a fifty year term with no increases. Our taxes went up. Our overhead and cost of insurance went up. We almost lost the place. Our acres of marsh land saved us.

An old guy, a musician, came in one day. I was at the shopping center in the adjacent hotel. The old guy said, 'Who owns that land out there?' I said I did. He had threadbare pants, shoes that were worn out, but he was a nice guy. He said, 'If I could sell that land,

how much would you want for it?' I said, 'A half million dollars.' He said, 'If I can sell it for you, will you give me a commission?' I said, 'Yeah, I'll give you $50,000.' He said, 'I'll be back.'

I called my partners and told them the story. They all said okay. About two weeks later the old man showed up and said, 'I got the land sold.' He drove in that afternoon followed by a station wagon. There was a guy driving the wagon and a man was in the back. A Mr. McCormick had broken his back. He laid on a board in the back seat. I went out and talked to him. He said, 'I need that land of yours. I'm in the construction business with Barbato and McCoba.' Their business name was McCoba Construction.

McCormick said, 'They're tearing down Scollay Square.' He had won the contract for moving all the

Business is
Booming at
Northgate
Shopping
Center in
Revere.
Behind the
center is what
was left of a
runway and the
seaplane base.

bricks. He asked, 'How much do you want for the land?' I said, 'Half a million dollars.' He agreed. I said, 'I need some money right now.' He gave me $50,000 in cash and then he left. Scollay Square was a famous place in Boston. They had the Old Howard Theater there, and famous restaurants. They tore them all down and built Government Center in their place.

My partners and I drew up the contract for the land. We had no mortgage. The little fellow came over to me and said, 'Can I have the $50,000 now?' I said, 'No way.' I said, 'I can give you $10,000.' I said he could have the balance when I got the rest of our money. By that time we were late on our mortgage. We had all kinds of troubles. This money salvaged everything.

Years after we opened Northgate, we took in a good management outfit. Allen Associates really turned things around. The place is beautiful. They're still managing the center and its 25 stores."

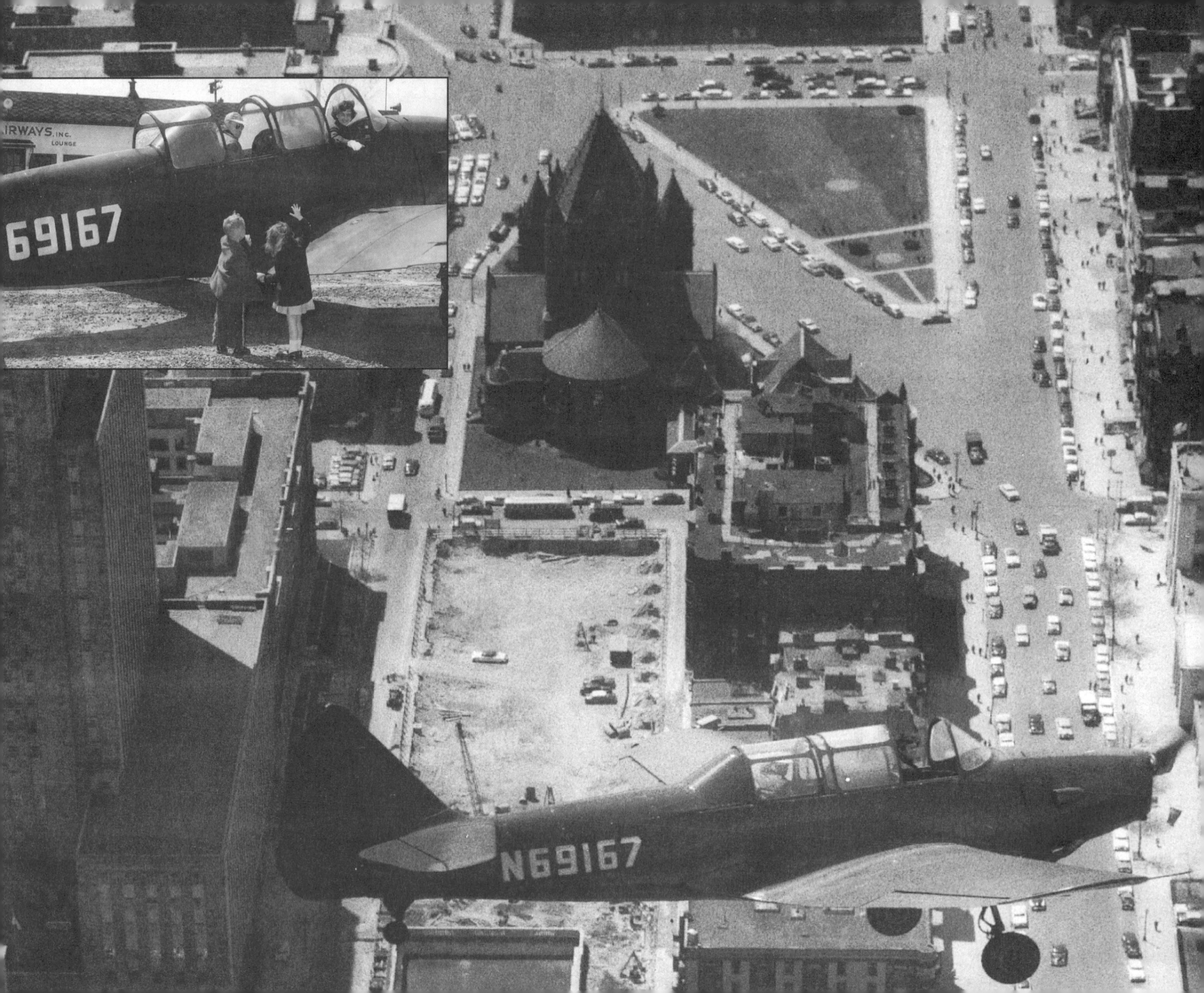
IRWAYS, INC.
LOUNGE
69167
N69167

CHAPTER SIX

Good Will:

"I WAS INVOLVED with the National Air Transportation Association in 1965. It was made up of pilots who flew for the airlines and did charter work for private individuals. They had headquarters in Washington, DC, and held annual meetings in different sections of the country. Colonel Ed Lyons from the Civil Air Patrol went to an annual meeting. He had an airport in Long Island near Republic Airport, a P-47 factory. The two airports had to avoid each other's air traffic constantly. Lyons and I were in the hierarchy of the NATA so when we'd go to a national meeting we'd room together. He asked me to get into the CAP. I was so busy at the time I said I wasn't interested. Lyons had given me books on CAP. At the next meeting we attended together, he asked me if I had read the books. I said no. The Civil Air Patrol started when World War Two began. There was a unit in each state, and a headquarters in Alabama. During the war, CAP did submarine patrol and border patrol. They had their own airplanes, not supplied by the government. It was done for the country by people who weren't going into the military. They formed an organization that could patrol the coastline against German submarines. They even bombed a submarine and sunk it. They flew all the borders of the United States. They took cadets from 12 to 18 years old and prepared them for going into the military. The cadets

IN 1956, BARBARA ELLIS FLIES OVER COPLEY SQUARE, BOSTON. JULIE WAS HER INSTRUCTOR. FLYING WAS TOUTED AS 'THE MODERN CAREER GIRL'S NEWEST HOBBY.'

were uniformed, and there were boys and girls as well as women and men in the hierarchy. The CAP cadets were learning to fly early. They could start lessons at 16 and by 17 they could have their private license.

About a month later I was at the Beverly Airport where we had our fixed base operation, Revere Aviation. I was looking out the window and I saw a military airplane taxi in. It was a T-34. From the window I saw two fellows step out in uniform. One was Ed Lyons and the other was an Air Force Colonel by the name of Bendix, whom I didn't know.

The Colonel talked to me for a while. He said, 'You're in the Civil Air Patrol now. You were appointed to be Commander of the Massachusetts wing.' I said, 'No way.' I was too busy and I had promised Florence that I wouldn't take on any extra responsibilities. Once

JULIE TOOK FOUR OF THE THUNDERBIRDS OUT TO MAP OBSTRUCTIONS BEFORE THEIR ANNUAL SHOW.

in a while I would actually go home at the end of the day. After I kept refusing the offer, he said, 'You've already been appointed a Lt. Colonel in the Mass wing.' I said, 'I don't care what you did, I'm not going.' Ed said, 'Let me talk to Florence.' He picked up the phone. Ed called his wife and asked her to call Florence to convince me. His wife called my wife and they talked a while. Florence called me about an hour later and said, 'Why don't you try it for a year?' That was in 1966.

If an airplane went down somewhere, the people who were sent to look for them were from the Civil Air Patrol. Their main mission was search and rescue. We found a lot of airplanes all over the United States. For example, when an airplane went down in Vermont, that wing was alerted immediately. If it went down in Montana, that wing was alerted. They'd search and try to find them, and most of the time we found them. Some dead, some alive. The Civil Air Patrol also took in youngsters and taught them all about aviation. They taught kids to respect authority and military ethics. Most of our cadets go to West Point and Annapolis, and to the Ship Academy down in Long Island.

I WAS THE WING COMMANDER of the Civil Air Patrol for Massachusetts. The region commander called me. He told me The Thunderbirds were having their annual show at Westover Air Force Base. The pilots wanted to make a low-level tour to map out any obstructions for the show. The jets they flew were too fast to pick up everything. They needed a slower airplane to fly them around. I said I'd fly them in my Aztec. The minute we got up in the air, one of the Thunderbirds asked if he could fly. I said, 'Go to it.' He flew around and the two others wrote down where any towers or obstructions were in their traffic route. They finished and the captain at the controls said I could land it. I landed my plane and they thanked me for the ride. They sent me a picture of their jets in flight signed by everyone.

The flight leader I met got killed in Ohio. He and his crew chief flew into a flock of seagulls. They ejected from the F16. His crew chief got out but he was killed. The rest of the guys got a new flight leader. They were out near Las Vegas doing a show when the flight leader started doing loops. He made a wrong turn and all four of the men were killed. After that time, the Air Force decided to cancel all their shows. The General in charge of the Thunderbirds said that there would never be another accident because the program would be canceled. They started the program again but they haven't lost a man since. The Blue Angels have, but they are Navy.

Flying in formation requires the pilot to fly beside another airplane and hold his altitude and distance from the other airplane. The Thunderbirds and The Blue Angels fly at tremendous speeds. They never home in on each other. They know how to hold the force of the airplane away from the other airplanes. That takes a lot of training. They have to see where the other airplane is all the time. They can't close in.

I was commander of the Massachusetts wing as Lt. Colonel. Then I was a Colonel for four years.

JULIE BECAME A REGION COMMANDER OF THE CIVIL AIR PATROL. NINE STATES AND 17,000 PEOPLE WERE IN HIS REGION.

Massachusetts CAP went from the second lowest ranking in the United States, 49th, to 29th in six months. After one year we were number two in the nation. At the end of four years the CAP had a national meeting for the region in Bridgeport, Connecticut. I went with the wing people and they gave me a special award for what I had done in Massachusetts.

Towards the end of the meeting, General Wilcox called me up to the stage. He asked me how I had been so successful. He said, 'You must of had a hard time.' I said yes. He asked what rank I was when I started. I said Lt. Colonel. He asked how I got that rank. I said, 'They started me at the top.' He said, 'There's no other way you can go except down.' Everybody laughed. He then said they were relieving me from duty because they were making me the commander for the region, for nine states. I was put on the spot but I still said, 'No way.' He said it was done. I said, 'No way, I won't take it.' I was only supposed to be in for one year. I had worked four years and now I was going to have to get involved with the region. Colonel Ed Lyons had recommended me for the job. He said, 'Take it, take it.'

The region was in pretty good shape. Finally, I weakened and I accepted. I held that job for eight years. The term was only supposed to be a couple of years. I took command of the region and we got to the top ranking for all the regions in the United States. Meanwhile I became a member of the national board of Civil Air Patrol at Maxwell Air Force Base, the headquarters, in Alabama. Florence was involved with the CAP, too. She organized the affairs, managed and planned events.

It was a chore to stick with because we had national meetings and we had nine wings. The northeast region was the biggest region in the United States. It included all of New England, New York, Pennsylvania and New Jersey. My office managed a total of 17,000 people. I had to go to each state in the region for their national meeting and communicate with them all the time. We had about 200 airplanes in the region. The government had certain surplus airplanes which we could requisition—T-34s, Beavers, Cessnas, L-19s. They didn't charge us anything for them, so we gave these planes to the different wings and the wings would try to take care of themselves. There was no money supplied by the government although there were active duty Air Force officers assigned to each wing.

I was so busy traveling that I couldn't keep up with everything. I told Florence that I would not take on any more work but she asked me to do it. I was president of the National Air Taxi Conference and I was working with the airlines hauling commuters for them. At the same time, I was a colonel in the Civil Air Patrol.

We did all the work for the government. They patrol the coast of Florida and different borders of the United States. The CAP watches for drugs coming in by plane or boat. Most of the equipment is surplus. The airplanes are supplied by the Air Force or the State itself. Vans are given to take cadets to different activities.

I met Zennin Hanson, president of Mack Trucks, when I was a wing commander for the CAP. Then I became the region commander and Pennsylvania was one of my states. We'd go to meetings together. We were both colonels. When I went into the army at 14, I rode in one of his company's trucks. I drove a Bulldog Mack truck when I worked for the soda company. Bulldog Macks had hard solid rubber tires, without tubes inside them. He gave me an award made with a miniature Bulldog Mack built in 1914.

One of the things we enjoyed was going to Cape Kennedy. We'd take a four-engine Air Force airplane and a group of people to see the firing of missiles to the moon. I saw the firing of the Apollo 16, on April 16, 1972. The grounds shook and the heavens split and this missile went up carrying three astronauts.

While I was there I saw Barry Goldwater. He was my squadron commander at the Air Force base in Wilmington, Delaware. When I knew him I was a 1st Lieutenant, and he was a Major. I didn't see him too much, I was in Newfoundland most of the time. I hadn't seen him since the war and we talked for a while. He had run for president against Lyndon Johnson in 1964 and didn't make it.

I made quite a few trips to Cape Kennedy. Once we took 18 nuns down from a Catholic college in Worcester. They had never seen anything like it. We had a four-engine Air Force airplane and we let each one of them fly a little bit. We landed and took them all through Cape Kennedy and then brought them home. They marveled at everything.

I took a group of college presidents from the New York area to Cape Kennedy. I told a president that I had never gone to high school. He said, 'Well, we can give you a high school certificate for all the background that you have.' I never went and got it.

NORTHEAST AIRLINES called me when they had someone wanting to go to an oddball place. Elizabeth Arden was going to Balsom Spa in New York, a racetrack

JULIE CAUGHT UP WITH BARRY GOLDWATER AT CAPE KENNEDY. GOLDWATER WAS HIS SQUADRON COMMANDER AT NEWCASTLE ARMY AIR BASE.

area. She had horses running there. I flew her and her secretary, both tall, good-looking women. The secretary came over and asked, 'Is that the airplane we are going in?' I said yes. She said, 'Miss Arden won't fly in it. It's too small.' I told her, 'You tell her I fly my wife and kids in that airplane.' She climbed into the airplane with her secretary. It was a good ride. I got them to Saratoga Race Track in an hour and a half. When we landed, the chauffeur drove up in a big car. She told him, 'James, give the man a $50 tip. It was a good ride.' I didn't tell her I owned the airplane.

I FLEW HAROLD STASSEN, the governor of Minnesota in a little Tri-pacer, to Concord, New Hampshire. He had run for president several times. That day, he met with Sherman Adams. Adams was Eisenhower's Chief of Staff and the Governor of New Hampshire.

Samuel Goldfine, a Boston businessman, was trying

from Goldfine, he fired Adams. He knew Goldfine expected something in return. A favor of some kind. Goldfine was thrown in the can down in Danbury, Connecticut, in a federal prison. He was in there quite a while. They fined him for trying to influence the President. I flew his secretary and his lawyer down to see him at the prison every two weeks. His lawyer, Robert Bennett, was trying to get him out.

Goldfine was selling 100 acres of land in East Boston. My partners and I bought it. We had the theater there already and we thought we might build something else on his land. I flew my plane out with Bill Hannan, who was shooting pictures for mapping purposes.

The city was designing an underground tunnel through Boston. I flew from the other side of the city, over Boston and came out over at our theater. When I saw these pictures I realized that the tunnel would end at the land we had bought from Goldfine. I told my partners. Sumner Redstone said, 'Well, we'll hold onto that land.'

The government took the land away from us, along with our theater. We settled for a good profit. It was split up between all the owners, and I got a pretty good buck. Goldfine eventually got out of the prison. Years later, Robert Bennett was President Clinton's defense lawyer.

IN 1959, NORTHEAST Airlines asked me to fly Admiral Arthur Radford from Logan Airport to Bar Harbor, Maine. Radford took a taxi to Nelson Rockefeller's house. A local man had built a plane in Bar Harbor and had it parked at the airport. Before we left

ADMIRAL RADFORD VISITS ROCKEFELLER. HE BECAME CHAIRMAN OF THE JOINT CHIEFS OF STAFF IN 1953. RIGHT: A TOUCH OF THE WRIGHT BROTHERS. TOO BAD IT DIDN'T FLY.

to make points with Adams and Eisenhower. He gave Adams a bolt of vicuna cloth to make a coat for the President. When Eisenhower found out that it was a gift for Boston, I showed Radford the funny-looking plane and paid the owner $1 to start it up. The wings were flapping on the plane. They'd drop down and it would jump in the air but it would never fly. On the flight back, I told Radford that I had a photo of Eisenhower in Paris in 1945. I sent him the photo after we landed and he delivered it to the President. Two weeks later, I got a letter from Eisenhower, thanking me for the picture.

I took the picture in Paris with my brownie box camera. The camera probably cost $10. It closed up into a box and when it was opened the bellows came out. I dropped it in the water in one of the lakes in Newfoundland. I was trying to take a picture out of the nose of my PBY. The plane rolled in a wave and I dropped the camera. As it was sinking, I hung down and grabbed it out of the water. I ruined my pictures in the bellows but I brought it out and I sent it to Kodak and they fixed it for me for nothing. They replaced the bellows and sent it back to me.

DICK BERENSON WAS good friends with Jack Kennedy. In the late 50s I went to a meeting in Worcester with my Bamboo Bomber. All the fixed based operators and the FAA Inspectors met. They talked about law and current issues. A man named Berwin Hyde was the head of the FAA office in Norwood. He was sitting beside me. Hyde said, 'By the way Julie, we're not going to renew your examiner's rating.' I said, 'What?' He said, 'Well, we decided we are going to cancel all the examiner ratings and do all the work ourselves.'

THE WHITE HOUSE

WASHINGTON

Gettysburg
August 20, 1959

Dear Mr. Goldman:

Admiral Radford has handed me the photograph you so kindly wanted me to have, taken at Orly Field in 1945. I am more than grateful for your thoughtfulness.

With best wishes,

Sincerely,

Dwight D Eisenhower

Mr. Julius Goldman,
25 Clyde Street,
Malden 48, Massachusetts.

The Federal Inspectors were taking away the private examiner's ratings. I said, 'Do you have anybody left?' He said, 'Yeah, one man, Charlie Mellie.' Charlie was a private examiner in Wiggins Airways in Norwood. I said, 'Why'd you keep him?' He said, 'Well, it's pretty busy there and we thought we'd keep him.' I said 'You got rid of everybody? You've got four or five inspectors and Charlie?' He said yes. I said 'I won't buy it.' He said 'What do you mean you won't buy it?' I said 'Just what I said, I won't buy it.'

I left the meeting and climbed in my airplane and flew back to Revere. If they wouldn't have kept that one man, I wouldn't have said a thing. It was late at night. I woke up my partner, Dick. I said, 'I lost my examiner's rating.' He asked why and I told him. I said 'Can we talk to anybody?' He said, 'Well, I'll call Jack.' At that time, Jack Kennedy was still a Senator. I came home, rested up and went back to the airport in the morning. I called the newspaper editor of the *Revere Journal*, Sid Curtis. I told him the whole story. He said, 'Well, why do you stand for it?' I said, 'I don't know who else to go to.' Sid said, 'I'll call the representative for this area,' who lived in Lawrence. He knew him pretty well.

Sid called me a couple of hours later. Sid said 'Get ready to go to Washington first thing in the morning.' The three of us were going to go and fight for my rating. That same day at 4 p.m., I got a call from Derwin Hyde. He said, 'Julie we've been looking over all the records and you've got a good record. We're going to renew your rating.' I said 'What made you change your mind?' He said 'Oh, just because of the records.'

I think Jack Kennedy got involved in Washington and called the headquarters of the FAA. They called Derwin Hyde and told him he better renew me. I had a little tri-pacer and the weather was bad that day. One of my students saw me get into my plane. He said, 'Where are you going?' I said, 'I'm going to Norwood to take a flight test.' My rating would expire that day and I had to go renew it. I squirreled my way into Norwood. I taxied in and shut the engine off. I went into the FAA office. They were in the same building as Wiggins Airways.

I walked in the door and they looked at me. A man said, 'What are you doing here?' I said 'I came for my flight test.' The guy said, 'Well, we don't fly in weather like this to give flight checks.' I said, 'I want my flight test. My expiration date is today, I want to renew today.' He said, 'You can't renew it unless you take a check ride.' I said 'Well, let's go.' He said, 'No, we don't fly like that. I guarantee you that we will renew it tomorrow.' I said okay and I flew back to Revere. The next day at 8 a.m. I was there and they renewed my rating.

I had my examiner's rating for 40 years. When I moved to Florida for half a year, so many people were waiting for me that I decided I should give up my flight test privileges. I asked the FAA to appoint two guys in my place. I recommended Rod Gorham and Bob Wade. The FAA said, 'Why should we put two men in your place?' I said, 'Well, there's a lot of work and they're both good.' They said, 'Well, we don't put two men in one job.' I said, 'There's other flight operations around so Wade could do the west side and Rod could do the east side.' They appointed both of them in my place. That ended that part of my career.

I WAS FLYING FOR THE newspapers and that's how I got to fly Bobby Kennedy. Laishke Ford was in the movie business. He knew my reputation around the city. He knew the Kennedys because they owned a theater chain. Laishke called me one day and said, 'Julie, Bobby Kennedy wants to go to the Cape.' Bobby was just a kid in college at the time. Jack Kennedy was a Massachusetts senator so I knew the Kennedys were famous. I was to meet Bobby at Logan Airport and fly him home to Hyannis, Massachusetts.

I went to meet him at 6 p.m. at Logan Airport. The weather was kind of poor but still flyable. I waited until 7:30 p.m. and he still hadn't shown up. I was ready to leave when Kennedy appeared with a friend. He came over and asked, 'Are you the fellow who is supposed to fly me?' I said yes. He said, 'Well, let's go.' I could smell the booze on him. Both of them had been drinking. I said, 'The weather has gone bad. You should have been here at six and we'd of had no problem.' He said 'Well, aren't you going to try?' They got in my Bamboo Bomber and we headed toward the Cape.

By the time I got over Nantasket, I was in the clouds. I had no flight plan for where I was going, and there were no towers. The Hyannis Airport was just a little square field that had been made with tar. There were a few old tin hangers on it. When I got up in the clouds I thought, 'This is no good. How am I going to find the place?' I tucked the nose down and came out of the clouds. At 900 feet I could see lights on the shore. I followed the shore and called Otis Air Force Base. I asked what the weather was like down in Hyannis. Otis had a 900 foot ceiling of visibility and three miles of rain. I thought there was no way of finding this field except by taking a heading. I went across the ocean.

By luck I found the field. It was bare except for a couple of lights on the side. The field didn't have a rotating beacon but I came across, saw the strip and landed. There was pretty good rain by then. I saw two tin hangers on one end, and I taxied over near them. A car had just pulled in the driveway at the edge of the

hangars. It was a Cadillac. Ethel Kennedy climbed out. Bobby Kennedy and his friend got out of the airplane, jumped in the car and left me there.

I was awfully mad. That wasn't the right thing to do. They should have asked if I wanted to stay overnight. There was heavy rain and no facilities, nothing. I was mad so I took off into the rain and headed toward Boston. North Boston was easy to find because of the homing beacons. When I was over Nantasket again, I called the Boston tower and asked what they had for visibility. I couldn't see Boston, I was flying by radio signals. A fellow at the tower recognized my voice. He said, 'What are you doing up there, Julie?' I said, 'I'm coming home from a charter trip. I want to get to my little field.' Revere Airways didn't have any beacons but it did have some lights that I had installed. The tower said to keep coming. They gave me steer and I landed in Revere.

A while later I got a call. Jack Kennedy had started running for president. Bobby was going to a Democratic convention in Springfield, Massachusetts. We took off from my field and I heard him talking in back. He was reading a speech and bemoaning the fact that he couldn't do what he wanted to do. He had to go speak for Jack. I thought that was pretty poor of the guy. His brother was running for President, and he was squawking that he had to go there and speak.

I heard him talk after the Bay of Pigs. Jack appointed him Attorney General in 1961. I heard him speak a few times and I thought to myself, he grew up. He sounded pretty good. Bobby Kennedy became a senator in New

York in 1964. He was running for president when he was killed in 1968.

I FLEW ATTORNEY GENERAL Edward

McCormack to Hyannis. I blew a tire on my twin-engine plane when landing at Hyannis. I brought the plane to a stop safely, but needed 250 feet to do it. I jacked up another airplane, took the wheel and tire off and put it on my plane for the flight back to Boston. I sent the borrowed tire back to Hyannis the next day.

It was August, 1962, when Attorney General Edward McCormack ran against Ted Kennedy for Senate. Teddy was 30 years old. When Jack Kennedy became President, his Massachusetts senate seat was open. Ted couldn't be a senator yet because he was only 28. Ben Smith was appointed to the seat for two years. Smith quit and the local politicians couldn't appoint Ted Kennedy now because it was an open field.

Eddie McCormack was attorney general to the state. He campaigned against Ted Kennedy. His uncle, John McCormack, was Speaker of the House of Representatives. In those days he was a real tough guy, he ran that House like a sledgehammer. Eddie McCormack was a good guy. I flew him everywhere during his campaign. I flew him out to the western part of the state and down the Cape. I'd land at different airports and he and Kennedy would talk to crowds of people holding signs and waving flags.

Eddie McCormack called me one time and asked if I could pick him up in North Adams. I rented a customer's airplane. It was a Piper Comanche airplane, it had low wings but big wing tanks on it. I took Shelia with me and we headed out at night. I saw there were a lot of people cheering for Ted Kennedy. Eddie McCormack had less of a crowd than the time before. He climbed into the plane. I told him, 'I had heard that the Kennedys would back you to be governor. Why don't you be governor?' He said, 'Julie, if you ever tell me that again I'll never fly with you. I'm going to make it.' He was a good guy and very smart.

When I took off, the weather was kind of poor. I found Brockton airport in the darkness. The airport was closed but Nat Trager owned it. It was a tiny field, only 1500 feet long. Trager was selling boats there. I called him up and asked if the airport was landable. He said yes. He didn't warn me about the bushes that had grown on the sides of the runways. There were no lights. I knew there was a highline on one end of the field. We were in a thunderstorm. It was raining like hell and flashing lightning. I was coming around over the field. Every time it flashed lightning I would look to see the wires under me. I landed over the wires into the open end of the field. Shelia was sitting in the front with me. Eddie McCormack sat in the back with one of his aides.

We made it onto the field. I was cutting bushes down with the wingtip tanks. I said, 'Goddamn Trager, he told me the field was okay and it isn't.' I stopped before I ran out of field. I said to Eddie, 'I can't taxi in toward the highway. You're going to have to walk in. You might get a little bit wet.' He and his aide climbed out of the plane and trudged off toward the highway.

The runway was so narrow I couldn't turn the plane around using the engine. I would've hit the bushes with the prop and I didn't want to do that. The plane was on tricycle gear. I didn't want to take off toward the wires. I got outside and sat on the tail of the airplane. Using my body weight, I turned the plane around. I got back in the plane and taxied out to the wires. When I got to the wires I turned it around again by sitting on the tail. I was drenched. Meanwhile Shelia was sitting inside and the engine was running all the time. We took off for home. That was Shelia's first plane ride in a thunderstorm.

A FELLOW WALKED INTO the office at Revere Airport one day. He said, 'My mother is dying and we want to get her to Yarmouth, Nova Scotia,' which is way up north. I said, 'I can't go because tomorrow is Yom Kippur,' an important Jewish holiday. He said, 'We have to get her there and you're the only one who can take her.' I said okay. I was fasting 24 hours for the holiday and didn't even have a glass of water. The next morning they brought her out to the airport in the ambulance on a stretcher. I said, 'Where's the nurse, where's the doctor?' The man said, 'No, my father and I will take care of her.' I said, 'Are you sure?' He said, 'We can handle her.' I said okay. The father and son sat in back. I had no copilot seat. The stretcher was lying on it. It was a clear day. By the time I got to Portland, Maine, it started to go overcast. I got up to Rockland near Bar Harbor and it was rough. The airplane was jumping and shaking.

I got close to Eastport, way up in Maine almost into Brunswick, and the son was sick on the floor. Then the father was sick. I looked back and the mother had her eyes closed. She was dribbling out of her mouth. I crossed the border into Brunswick and landed. The customs guy climbed up on the wing and opened the door. He said, 'Oh my God.' The smell was terrible. He fell off the wing and said, 'You're cleared.' He wrote the clearance and I took a blanket and covered the mess.

I headed across the Bay of Fundy, across to Nova Scotia, and it started to snow. I was going all the way up north, near Newfoundland. The situation was no good. I called the Navy base in Halifax. The Canadians had a base there, called Dartmouth. I told them I had a dying woman aboard, and I had to land. I asked, 'Can I clear and get into the navy base?' They said yes. I went across and landed. The base was only five miles from Halifax. They got the woman out and put her in an ambulance while I cleaned up the airplane. When I finished I said, 'There's a road that runs right to Sydney from here. I'm going home.' That was on Yom Kippur. I got home and I had a roaring headache. I paid my atonement that day.

Occasionally people would leave instructions in their will to scatter their ashes over the ocean or over the mountains. Only one person from the family would usually go with me. Sometimes I would do it by myself. The first ashes I spread were from a fellow who had a little boat and used to swim out in the ocean. Before he died he asked to have his ashes spread over Boston Harbor. I opened the window of my plane. I had the urn facing forward. Inside the urns there were ashes and small

THE ANDREA DORIA SANK SHORTLY AFTER THIS PICTURE WAS TAKEN BY OLLIE NOONAN.

stones used in cremations. The cremating stones were left in the ashes. When I dumped out the urn the stones came out and hit me in the face. All the ashes came out too, all over me. That was a mess. I had to have the urn facing backwards so the air would suck the ashes out.

I spread ashes over the mountains in New Hampshire, Maine, and Vermont. I usually flew over specific areas like North Conway, or Mount Washington. The ocean requests were usually over harbors.

IN JULY OF 1956, I was the first plane out to the *Andrea Doria* when it was sinking. *The Boston American* and the *Daily Record*, the *Boston Globe* and the *United Press* hired me to fly photographers out and get pictures of the collision. I told them I didn't think I could find it. I took three photographers and climbed out on top of the fog and out to sea.

I was given a location of 049 degrees, 50 miles south of Nantucket. I saw a ship chimney sticking out of the fog and I descended down. I was over the *Il de France*. The ship was taking a load of people to shore. I circled it a couple times and descended down into the fog. I flew along the water trail of the *Il de France* and saw the Stockholm. Her hood was crushed but she was pushing her way toward land. I circled the boat twice and the photographers went to work. I followed its wake and we saw the *Andrea Doria's* lights. The boat was listing a little bit. Ollie and the others took pictures of the hole in the side. The hole was bigger than my airplane. It was about 30 feet wide.

I flew to Nantucket and dropped off the three

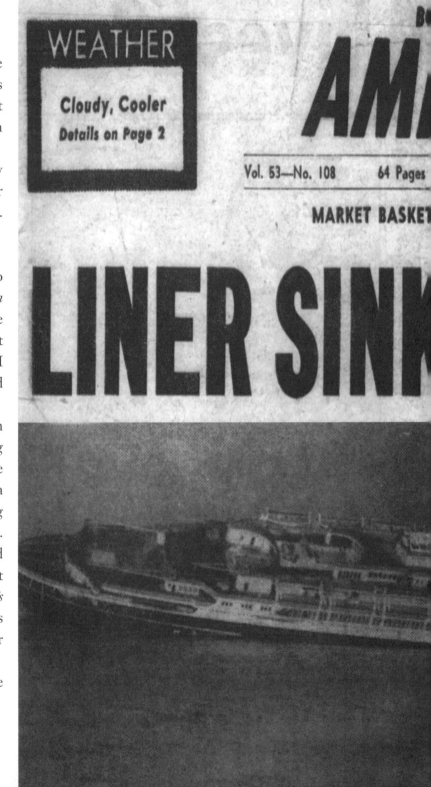

WEATHER

Cloudy, Cooler
Details on Page 2

Vol. 53—No. 108   64 Pages

MARKET BASKET

LINER SINK

photographers. They stayed there to be close to the action while I flew the pictures to Boston. The papers had a taxi waiting to get the pictures to the lab. I flew back to Nantucket, picked up Ollie Noonan and the other two and flew back to the scene. I couldn't find the boat. There were rescue boats with people in them, but the Andrea Doria had sunk. Four people died.

THE CIVIL AIR PATROL wanted me to go to Austria with General Wilcox, a three-star general in Alabama, for the planning of an annual event. I said I wouldn't go. The General called me. He said, 'Don't you realize the honor?' I said I did but I didn't. I was too busy running conferences in nine states. I said, 'I don't have time for that.' This trip was to plan for a gathering of Civil Air Patrol cadets representing all the countries. Once a year, the International Air Cadet Exchange met somewhere from all over the world. I was to represent CAP and the General would represent the Air Force with CAP.

I got a congressional report, which updated me on events. I saw a picture in the book of a boy in front of an ancient wall. I asked someone, 'Where is that?' I was told it was Israel. The wailing wall is where the Jews asked God to help them. It is the holiest of places in Israel. This wall is the only thing that's left that Jews can see from the original temple built by King Solomon. In 1967, the Israelis fought the Six Day War. They took the wailing wall back from the Arabs. The buildings around the wailing wall were removed and the area in front of it was levelled into a large paved open space for prayer.

I had given money to support Israel every year, and I said I'd like to go there. The CAP people said, 'Well, you turned it down.' That's when I realized there was more to this thing than just being involved with the Austrians and the Germans. I called the General. He said, 'You said you understood the honor we were giving you. You didn't understand it because you turned it down. What makes you think that we're going to give you a rain check?' He hung up.

Two months later I got a call from Maxwell Air Force Base. The secretary said, 'Colonel Goldman, you'll be receiving tickets from the airlines to go to Washington. You'll get briefings in Washington and then you're going to go out to Germany and on to Israel.' We had cadets from the Civil Air Patrol and I was their escort. Twelve cadets were to travel with me from the United States. A woman, Ione Hamman, was a commander of a squadron of female cadets. She was in charge of Jackie Richardson from Chicago and Laneise Yamamoto from Hawaii.

The other ten cadets were from different parts of the States. We were sent on a four-engine jet from Washington to Rhein Main, Germany. I was a full colonel and they put me in command of the folks on that jet. We took off from Washington and got to Germany. Hundreds of cadets from all over the world came in to Rhein Main. Representatives from 20 nations were flown in from different countries—the States, Canada, Switzerland. They had a big blowout the night after we arrived with three or four hundred cadets in one ballroom. In the morning we were to go to the assigned countries. I was assigned to Israel with 12 people, Ione Hamman and myself.

We got up in the morning at 5 a.m., and got ready for the day. I was assigned to an old airplane that the French had built called a Nord Atlas. It was a twin-engine airplane like a troup carrier. The bus took us to Operations. I walked in and checked on the Operations Board. I saw, 'J. Goldman, Nord Atlas,' and a list of the people who were going with me. There were eight new names on my list. I thought they made a mistake. I walked up to the Operations desk and said, 'You made a mistake, you have eight more people with me.' 'No, we didn't make a mistake. Some countries can't afford to send escorts for their cadets. We assigned additional people to you.' I said, 'I can't take them. They won't be under my orders.' He said, 'They're going to be under your orders.' I said, 'What if they don't accept them?' He said, 'Well, then you can send them home.' I said, 'How am I going to do that?' He said, 'You call the ambassador for their country.'

I took the other cadets. There were two guys from Norway, two from Holland, two from England and two from Canada. We walked over to the Nord Atlas. Our bags were packed for three or four weeks. I started to go up the steps into the airplane and a civilian stopped me. He said, 'Go back to your bag.' He said, 'All you people, open your bags.' There were 21 of us out there opening bags. I said, 'Why do we have to, we just packed them.' He said, 'Did you have that bag in your hand since you packed it?' I said, 'No, I had breakfast.' Just then, the Lt. Colonel of the Israeli Air Force looked out from the plane. He said, 'Who's the commander here?' I said, 'I am, Colonel Goldman.' He shook my hand and said, 'I'm Colonel Itzhak Barkin. You do what the civilian says. He's security for your safety and for our safety.' The bags were all open. Our stuff was all over the ramp. We finally got them closed up, and boarded the plane.

There were 12 men on the airplane already. I said, 'What are these guys doing in here?' Colonel Bakkan said, 'They are spare crews.' I said, 'Well, what are they doing here?' He said, 'We don't have the money you have so we are using this as a training mission. We trade off parts of the flight to each man.' There were three men in the cockpit. The others swapped off. The airplane was loaded in the belly with a ton of stuff that they bought in Germany. We took off and the plane was so heavy that we couldn't go over the Alps. We had to go around them.

The first stop we made was in Brandizzo, Italy. Brandizzo is in the elbow of Italy. We landed there and Barkin said, 'You can have anything you want, drinks, or food. We'll pay for everything.' I said, 'No way, they are not drinking under my tutelage.' One of the English kids said, 'They allow us to drink.' I said, 'I don't care what they allow you to do in England, you are not drinking under my command.' We went to some nice restaurants. I didn't know it but the English boys were out drinking. The American boys were watching them and some of them started to drink. I found out and I gave them hell. I threatened to send them home. We left Italy and flew to Israel non-stop. We weren't flying too high. There were clamshell doors in the back of the plane that opened up for parachuters. The doors didn't close too well. I was sitting in the tail, and I could look out at the ocean.

Just before sunset we landed in Tel Aviv at Lod Airport. The kids were dressed in torn pants and overalls. I was the first one out and was greeted by a whole row of people. A man in uniform stepped forward. He said, 'I'm Major Zinger.' He shook my hand and started introducing me to all the hierarchy from around the city. The kids started trouping out in the torn pants. The Major said, 'Get them back in, they're on national television.' They asked me, 'How do you like Israel?' and I had just landed. I said, 'It's great.' We got the kids squared away and dressed up. The cadets came out and were introduced. The Israelis took us everywhere and got pictures of our trip. The first night we watched ourselves on television and I was saying 'It's a wonderful country,' and I'd only been there one day.

We went from tip to tip and border to border. We went to Elat which is near the ocean. It was a hot ride across the desert. One of the Dutchmen ran into the hotel. He changed into shorts, ran out to the swimming pool and dove in. There was no water in the pool. He knocked his teeth out and was all banged up. We saw Ramat Gan and Jerusalem and the Wailing Wall. We went on trips to Nazareth, Jericho, Mt. Carmel, and the Sea of Galilee among others.

The worst kid that I had was 18. He was a Jewish kid from Miami. All the cadets had to be of a certain age and reach certain marks to qualify for this trip. They picked him from Florida because he said that he could speak

He was about four chairs from me. I said, 'Get up and say something in Hebrew.' He looked up at me and said, 'I don't remember anything from my bar mitzvah.' He had lied on his application. Everyone was embarrassed. They asked me to say a couple of words. Then Major Zinger spoke in Hebrew to the group.

We had affairs in every city in town. They took care of us. The cadets were brought by bus to their host's home at day's end. They assigned each cadet to a separate home in Tel Aviv and the surrounding area. The bus dropped everyone off. In the morning, the cadets were to have breakfast with their host families. The bus picked up the kids at set times. Each cadet brought their dirty clothes to the bus in the morning. The Israeli Air Force would clean their clothes and have them back at night.

They wanted me to stay in a hotel. I said, 'I want to see how the people live here.' They said, 'No, you go to a hotel.' I said, 'Okay, but Ms. Hamman is going to the hotel, too.' They put us in separate private rooms. We met at breakfast together each day. The Israeli breakfasts were marvelous. We had olives and eggs and a breakfast feast.

Major Zinger was in charge of the cadets from Israel. He liked to come over to my hotel and have breakfast. We would be the last ones the bus would pick up, so we always started from our hotel. It was hot in the desert and the bus wasn't air conditioned. They'd stop and open up a crate and each of us had to drink a bottle of soda every 30 minutes. Ione got so hot in the back of the bus she was lying on her back panting. The trip was great.

Hebrew. He was given a bar mitzvah. He probably learned Hebrew but he didn't know how to speak it.

We went to the training base in Israel, and they treated us like gods. They had cadets and officers and us at a big ring of tables. The base commander got up. I didn't know him. He said, 'We have a surprise for all you cadets. We have a cadet here who can speak Hebrew.' I turned to him. He was turning red. I walked over to him.

They took us everywhere. Mayors fed us. I had to eat a gefilte fish at every stop.

The last night, the Israelis had a formal ball on the fire station roof. The cadets were to give gifts to their host parents and say goodbye. That afternoon, my non-Hebrew-speaking cadet from Miami said, 'Colonel Goldman, I don't have any money.' I said, 'What do you mean? You were told to leave the States with $50 spending money for the things you might need.' He said, 'Well, I used it up. I don't have any money and I have to give my host parents a gift.' I asked, 'What do you want to get them?' He said, 'I know they'd like a couple bottles of booze and some cigarettes.' I said, 'Okay, you'll have to pay me back. I'll get it for you.'

The two Norwegians also came up to me. I called them Little Norway and Big Norway, I couldn't pronounce their names. They said, 'We're ashamed to tell you we have no money for gifts.' I said, 'Why not?' When we stayed in Germany, we were housed in an American base. At 5 a.m. the cadets left all their stuff on their bunk, and took showers. The Norwegians came out of the showers and somebody had stolen their money. They were ashamed to ask for money but they had no choice. I said I'd get a couple of cartons of cigarettes in our embassy. Cigarettes were $20 a carton on the street but in the embassy it was only $1.80 a carton.

The Israelis treated us wonderfully. The ball at the firehouse started great. All the families came with their hosted cadet and everyone gave gifts. The Israelis gave each of us a photo album of our trip. At the end of the evening, the Israelis were talking Hebrew. I could tell they were talking about me. They were shaking their heads.

The next morning, we were out at Lod Airport, now called Ben Gurian. The fire chief who handled all the housing, Colonel Elkes, came over to me. He said, 'Goldman, we have a *teef* in our midst.' I said, 'What do you mean, a crook?' He said yes. I said, 'Who is it?' He said, 'That boy over there (the Miami cadet).' I said, 'Oh, I don't believe it.' He told me his host family wasn't happy. He came in from every trip dirty. He went to sleep in dirty clothes, not even taking his boots off. He borrowed money from his host family to go out to night clubs. He said he'd give it back to them before he left. When he didn't pay them back, he said, 'I'll give you cigarettes and booze to pay for it.' He was paying them off with the presents I bought for them.

I called him over. I told him I knew what he had done. He denied everything. I told Elkes that I'd talk to him on the airplane and straighten him out. He said, 'Oh no, you can't do that. He will say that we're liars. You'll have nobody to talk to after that and you'll never see us again.' I said, 'Bring him over here.' Ione Hamman was behind me. She said, 'Let me kill him, I want to kill him.' She was a very nice lady but she lost it. I had the kid write an apology. It said he promised to send all the family's money back to them. Then he would pay me off. We all went home. He returned the family's money but I never got a penny. I got a letter from the host family three weeks later thanking me. Later this kid wanted me to recommend him for ROTC. I said no and wrote up a

OPPOSITE: IN 1970, JULIE REPRESENTED THE U. S. CAP AT THE INTERNATIONAL AIR CADET EXCHANGE PROGRAM. HIS GROUP TRAVELED TO GERMANY, ITALY, AND THEN GOT A GRAND TOUR OF ISRAEL.

bad report. He was the only Jewish kid I had ever, and he was a louse.

DAVID ZINGER WAS THE first Major I met who was in charge of CAP for the Israeli government. He went to Germany with some cadets. He came back, reported in and had a heart attack in the General's office. I flew over to Israel to see him in the hospital and they wouldn't let me in. It was a military hospital. Somebody told them who I was so they put a gown on me and I went in. He had been unconscious the whole time he was there. I looked at him and said, 'David, this is Julie. I came from the States to see you.' He hadn't moved and all of a sudden his foot went up in the air. The nurses were surprised. He knew I was talking but he was still bad. He died about six months later.

MASSACHUSETTS AVIATION Trades was a State association I got involved with when I first opened my airport. In a couple years they made me president of all fixed-based operators. Every airport in Massachusetts had a fixed base operator on it. Hanscom had two or three. We ran little airports with flight schools. We were an association for pilots who ran charter flights. Because I was president for the State, I was invited to go to the annual meetings of the National Aviation Trades Association for the whole country. In 1950 I became a member of the NATA. They made me the vice president of the east at a national meeting in 1953. I was interested in anything to do with charter.

Wiggins Airways had a big operation at Norwood Airport, and a second one at Logan Airport. Joe Garside was the president. He wanted to form a national outfit called the National Air Taxi Conference. He started an association for anybody who wanted to just run charters. They would be under air taxi. I was interested so I joined. Not all airports had taxi operations going. In 1960, I was voted president of the NATC. We got involved with the airlines and tried to get them to give us space. Space was tough to get at every airport. Air taxi planes flew in passengers to the commercial airlines. It was quicker for both parties if we could land right at the United, TWA, Pan-American or Eastern terminals.

United Airlines recognized us first. Bob Shippey was the head of Interline Relations, United's association for all air taxi operations. He gave us $10,000 for a logo and printed books and gave us slots to park at different airports. He became a good friend of mine. Bob believed in building up air taxi. United could get a lot of passengers from towns and small airports with the help of air taxi. Other airlines saw that they were losing out because we were bringing our passengers to United. Then American got involved. TWA got into the act. Soon we had slots for each major airline at many airports. We fed the airlines passengers and were treated well in return.

My lawyer, George Rittenburg, had a niece who went from Ft. Lauderdale over to the Bahamas in a British De Havilland Dove. The plane crashed and everybody was killed. There was no life insurance for the passengers. My lawyer called me up and said, 'How can that be?' I said, 'The law doesn't cover that.' While I was president of the NATC, I put in a bill to the Civil Aeronautics Board. It stated that every airplane flying under air taxi conditions had to have insurance. Every plane had to have a minimum of $75,000 in insurance per seat. We got the bill passed at $50,000 a seat. The new law covered the United States and Hawaii.

WE WANTED AIR MAIL to go to towns that the post office drivers couldn't reach. J. Edward Day was an Irishman from Revere who learned how to fly at my airport. He was a supporter of presidential candidate John Kennedy. When Kennedy was elected President, Day became postmaster general for the United States. I called him up in Washington, D. C. He said, 'Come on down. I want you to go to the White House and meet Kennedy.' I had to say no because I was too busy. The next time I went into Washington for a meeting I called him. I told him I wanted to talk to him about air mail. I knew that some towns were difficult to get to by truck or train or boat. He said, 'I'll have a meeting set up with my staff. Bring a couple of your staff with you.'

A girl worked for me full-time in my National Air Taxi office in Washington. She'd run the office and call me in Boston with news. My staff and I went to the meeting in Washington. Several people were sitting around a long table. I sat down with my staff and told the post office people what I wanted to do. They said, 'We have a place in Delaware that we can't get the mail to except overnight by train.' The NATC got the first

contract on a Bonanza airplane, and flew the mail directly over the lake and the railroads. It now took an hour and twenty minutes to deliver the mail there. That started the air taxi air mail business which continues today. Small airplanes fly charter and deliver the mail to towns that are hard to reach. Before that partnership between the Postal Service and the NATC, all the air mail went by commercial airline.

The airlines took mail all over the country and overseas but not to towns with only ten bags of mail. If it wasn't on the regular airline route, it went by train or by bus. When I used to drive a truck, Greyhound buses

THE NATC PROVIDED AIR AMBULANCE SERVICES. JULIE WAS A VICE PRESIDENT IN 1958.

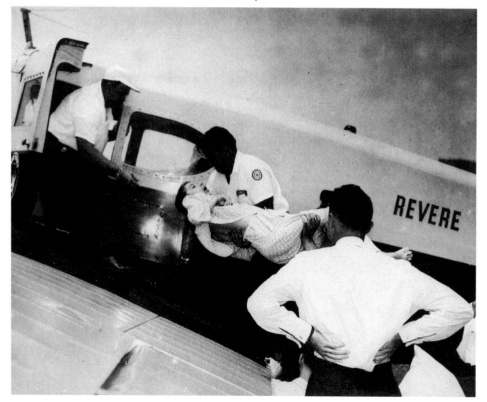

carried mail as well as passengers from Boston to Lewiston, Maine. Lindbergh used to fly the mail.

Al Lechshied, who gave me my first ride in an airplane, flew the mail in an open airplane. He hit a mountain one night and survived. He couldn't tell in the dark how high he was off the side of the mountain. He just hung on all night, and when it got light he said the ground was a couple of feet under him. He was up there for five or six hours.

I COULD FLY FREE ANYWHERE as a result of my work with the airlines. They'd ask me if I'd like to go to Hawaii, and would send Florence and me first class tickets. I'd say, 'I want my wife to go,' and Bob Shippey at United would take care of it. TWA had the Interline program. If you worked for the airlines you could get certain privileges with other airlines. I looked in the Interline book and I saw I could go around the world for $203. I said, 'Florence, do you want to go? Just the two of us?' She said yes and we went. We went with Ed and Pearl Lyons. Ed was in charge of the Civil Air Patrol in New York at that time. The four of us went around the world for $203 each. We paid for our lodging. We stayed as long as we wanted, anywhere. Then we called TWA when we wanted to go to our next stop.

It was romantic being with Florence. We met Ed and Pearl in Paris. Ed had married Pearl in Paris. Her family owned a big jewelry business there. They took us everywhere, underground and overground. Some of the French restaurants were three stories underground.

From Paris we went to Israel. We stopped in a hotel not far from the airport. I called some of the friends I had met the year before. I told them I was in town and they were ecstatic. They didn't believe I was going to come back. My Israeli friends said, 'You can't stay in that hotel you booked. You have to stay in a hotel on the ocean.' They made arrangements for us to stay at a classy hotel.

We were in the airport hotel lobby waiting for my friends. A small guy was walking a little dog. I kept looking at him and noticed everybody gave him a wide berth. Finally when we went outside, I asked, 'Who was that guy?' It was Meyer Lansky. He was a well-known gangster from Florida who left the States to get away from our government. He owned a lot of properties in Havana and the government was trying to get him back into the States. The Israelis told him he had to leave because they had a good relationship with the U. S. He returned home and the U. S. government didn't do anything to him.

We went to the hotel on the ocean and we saw all the people that I had met the year before. It was wonderful. My friends in Israel treated us like we were gods. From Israel, we went to Bombay. Our plane tickets were dependent on space available. Every time the four of us called TWA for a flight, they made sure we got on. We got to Bombay and we hated it. It was dirty everywhere. The very next day we left for Bangkok, which was beautiful. There were some dirty places, but it wasn't everywhere. We went on 15 foot boats with big outboard motors, and traveled up the rivers.

We were warned not to have any food from street vendors because it was filthy. We ate in the restaurants only. We hired a cab to take us places. I sat in front and Florence and Ed and Pearl were in the back. On one cab ride, as soon as I sat down, the driver showed me a little book. In it was a picture of a bare little girl. He said, 'Very cheap, very cheap.' I said, 'What are you, crazy?' Florence asked, 'What is he saying?' I said, 'Nothing,' and threw the picture out of the window.

We went inland one day. We rode the elephants and had a good time. But the people were filthy. They were swimming in the water around their shack on the river. The river was also their toilet. We didn't even dare to buy a banana. We stayed at a nice hotel, and it was there that I was sick for three days. I think it was the food. Next, we went to Hong Kong and that was beautiful. Florence had some rings made there. We brought the stones but she wanted settings made. The settings were solid gold and only cost $20. In Hong Kong, I had an Air Force uniform made with two pairs of pants for $35. It was ready overnight. Chukkerboots were $18 made with korofam, shiny black boot material. Florence had dresses made with pearls all over for $35. I bought a camera lens for a fellow pilot. He wanted a 500mm Nikon lens. It cost $2000 in the States, but I bought it for $350 in Hong Kong. I bought time-date watches for every one of my kids.

We went from Hong Kong to Guam, a flat little island. It was hot as hell. From Guam we went to Hawaii. When we took off from Guam in a 707, it was loaded to

JULIE, STILL ACTIVE IN THE QUIET BIRDMEN, BECAME FRIENDS WITH, FROM LEFT TO RIGHT: HARVEY SHULMAN (NOW A SENIOR CAPTIAN FOR DELTA AIRLINES), DOC STARK DRAPER, JULIE, FRANK KELLERHEA, CARMEN PERROTTI, SR.

the hilt. I never thought we were going to get off the ground. It looked as if we left the ground and were going into the ocean. We didn't clear the ocean by ten feet. About half way to Hawaii there was a commotion in the plane. A man collapsed because the air conditioning in the 707 wasn't working. There was a big woman sitting beside me. She was a nurse and she was pregnant. The crew got the guy by the heels and dragged him into the back of the cabin. The nurse went and helped the man. He had a small heart attack but finally he came to.

We got to Hawaii and we'd been there a few times so we knew our way around. We were leaving the hotel in Hawaii for our trip home. The suitcases were filled with things from every stop we made around the world. I said, 'Okay, everything's packed.' Florence said, 'No, I've got something in the drawer.' I said, 'What do you mean you've got something in the drawer? I told you not to buy anything else.' I had no place to put it. I grabbed her big brown hat box. I took off the top. There was a wig in there with a styrofoam head inside. I tossed them out the window and filled the hat box up with her stuff. She got so mad at me. We had left home with one suitcase. We now had 12 suitcases and the hat box.

ON ONE OF MY MANY TRIPS to Israel, a friend of mine needed some booze and cigarettes. I used to shop at the Embassy. As a CAP officer I had a special pass. I bought a few things at the commissary. I carried a shoulder bag all the time which held my passport and my money. I came out of the embassy through the back door. I asked Colonel Giora Oren to meet me there. The back of the Embassy in Tel Aviv faces the ocean. I walked out, and waited on a bench. Giora pulled up. I took the shopping bag off the bench and put it in the truck.

We were riding down the main drag when I realized I had left my shoulder bag on the bench. Giora tried to race back but the traffic was heavy. We finally got onto the shore road. When we got to the Embassy I saw my bag, sitting on the bench. I started to run to the bench and all hell broke loose. People from the embassy had called the authorities. They thought the bag may have had dynamite in it. I heard sirens coming. The embassy people were yelling, 'Go away, go away,' because they didn't know it was mine. I grabbed the bag, jumped back in the truck and we took off. I heard the sirens but they never caught up to us. I went back to the embassy later. I told them, 'That was my bag, there was nothing in it except my passport and money.'

IN ISRAEL, the Israeli kids called me grandfather. Florence and I had flown to Israel one year just before Labor Day weekend. When we arrived, Florence went to our friend's home. I was flown to the desert with the commander of the Israeli Air Force. We were outside and it was very hot. The food we ate was sitting out in the sun. I was sweating and didn't feel good.

The commander introduced me to his deputy commander, Avenue Ben Noon. He was going to Boston that night where he was a student at Harvard Business School. The commander asked me if I could do Avenue a favor or two when I got home.

When I got back to the States I got in touch with Avenue Ben Noon at Harvard. He was in school with a man named Smith who was a vice president for Buick. I took them flying one day with a colonel from the Thunderbirds. They loved it. Smith talked of GM business. I owned a Buick at the time and said I had a few complaints. Two days later, I got a call to bring my car into the dealership. I went in and met six engineers who checked the car from top to bottom and took care of my complaints. I never had another problem with it."

CHAPTER SEVEN

A Look Back:

"**I LOVED TO FLY** in good weather. But during the war, good weather was rare. If it wasn't snowing it was raining. If it wasn't raining, there was fog. In some planes you had to have a parachute. I never had to jump out. But I had my own chute and had to have it repacked every nine days. They didn't call for chutes in a four-engine unless I was traveling from a pursuit flight to another pursuit flight. If you were the pilot, you were to stay with the plane until you went in.

All the pursuit planes are retired. They're antiques now, but they were the best fighter planes we had. The P-47, P-49, P-51, P-40—these were the majors in World War Two. The P-47 was built by Republic and nicknamed the Thunderbolt. They used to paint a shark's mouth under the nose of a P-40. During the war, the fastest I ever flew was 250 miles per hour. Now planes can go 1000 miles an hour on a straightaway.

I took Marlene and Myrna flying when they were old enough, but they weren't interested in it. Shelia was interested. She wanted to know everything. Now she works for the FAA in the New England region.

I don't enjoy being the passenger but I occasionally have to be. My first flying license was issued in 1935. I renew my instructor's license every two years. I have to go with an examiner to renew it, and I used to be an examiner for forty-some odd years.

FLORENCE TRUSTED JULIE, BUT STILL PREFERRED THE BACK SEAT. HERE THEY PREPARE TO TAKE OFF FROM BEVERLY, MASSACHUSETTS.

I WAS COMMANDER for nine states in the Civil Air Patrol. We had 17,000 people in the region. I've been in it since 1966. I'm still a member and I keep giving donations for training the kids. I'm a charter member of the Aircraft Owners Pilot's Association. They have 367,000 members today. My membership number is 534.

I became very friendly with the Israeli Air Force. I had been to Israel for the Civil Air Patrol twice. I met commanders of the Air Force, many Colonels and lower ranks. We became personal friends. I went there 24 times. I flew on El-Al, an Israeli airline. A pilot from South Africa came to Israel. He flew in their Air Force and became a pilot for El-Al. He heard I was aboard his flight one time so he called me up. I went upstairs in this big 747. Florence came up with me and we sat up in the lounge area.

The captain called me into the cockpit. The pilot, the copilot and the engineer were seated. The captain said, 'Do you want to fly a little bit?' I said, 'I'm happy here.' He said, 'No, sit up in the seat.' I said, 'It's too crowded up here.' He said, 'No, get in the seat, get in the seat.' He had the copilot get out and he unlocked the autopilot. He let me hand fly it for a while. The plane was full with 350 passengers. I couldn't see the wing it was so far back. It was the biggest plane I had ever flown. I rolled it a little bit to get the feel. After I did that a few times he said, 'Stop, you're making my passengers sick.' I kept it on course. My friends took movies of me landing the airplane in Israel. They treated me well.

Sitting in the cockpit changes everything. You can see what's going on. You're not sitting down trying to sleep among people who are pushing you around. I came home to Boston on El-Al one time. I called my airport from the cockpit and told them what time I'd be home. The pilot said, 'Don't tell them the airplane number.' I used my own airplane number. My operator said, 'Where are you? I said, 'I'm overhead,' I didn't tell him I was 30,000 feet overhead.

NAT TRAGER BOUGHT a flight school and airport in Brockton after the war. Another fellow had the airport going then and they became partners. He was a good-looking guy. He's still married and he and his wife live in Florida. He's still active in aviation. He sold the flight school. There's a shopping center there now.

WHEN I LOOK BACK at those early days I will always remember Doc Sagansky with appreciation and respect. Doc always treated me fairly. When I wanted to pay him back some money from my airport loan I couldn't find him. He was always in Florida. Florence and I were in Florida and I looked for him one day. I found him on the main street in Miami, walking with his wife. In my pocket, I had $10,000 in bills. I caught up to him and said, 'Doc, I've been looking for you. I want to give you some of the money that I owe you.' He said, 'Oh, don't bother me with that stuff.' I stuffed it into his hand. He was such a good guy. He trusted me. He knew he would get the money.

SUMNER REDSTONE AND I were pretty good friends. Even to this day, I still get a pass to all his theaters. The pass is the first of any he gave out. It's like a credit card admission for two free seats at any Showcase Cinemas any time. The open air theaters are not around anymore, but he built four or five hundred of the Showcase Cinemas. He invited Florence and me to this great big affair in California at Universal Studios. Sumner was president of the Movie Theaters Association by then.

In October of 1965, Universal Studios had a big party. It was called the Theatre Owners of America Dinner Dance. Universal was marketing its movies to the owners. They had stuntmen perform, jumping off a roof into the street. Most of the movies were done in one huge studio. We toured the building and it was very interesting. Max Factor did Florence's hair and she was given sample bags of makeup to take home. She had a great time. They had huge halls filled with tables and at each table was a movie star, sitting among the guests. We were there three days and stayed at the Hollywood Roosevelt Hotel.

WHEN I MOVED my operation to Beverly Municipal Airport, Revere Aviation was one of three airport operators. Sylvania had a hangar there and North Atlantic had an operation. We leased space because we couldn't buy it. The city owned the land. I ran it for quite a few years and then I bought a place in Florida for Florence and me. My daughter Shelia came in and ran the company.

JULIE WITH HIS DAUGHTER, SHELIA BAUER. SHELIA TOOK OVER THE BEVERLY OPERATION, REVERE AVIATION, WHEN HER DAD RETIRED.

I worked hard until I was 70 and then I eased up a little. I gave Shelia an airplane that I bought for her. She still has it. Shelia wanted to sell the fixed based operation in 1989 when her partner died. Between Revere Airport in Revere and its second generation, Revere Aviation in Beverly, we enjoyed a life span of 43 years of aviation in New England. Shelia went to work for the Federal Aviation Administration and is the manager of their aviation education programs. We're still a team, as I usually help out at many of her outreach events for youngsters. My legacy to her is my love for aviation which she passes on to all the young people in her FAA education events.

Shelia is also an advisor to Aero Club of New England. Her husband, Jeff, is a director. Cabot Awards are given out every year. Dr. Draper, Crocker

Snow, Russ Boardman and Johnnie Polando all received awards. I was a director, and I'm now an advisor. I have many wonderful friends who are A. C. O. N. E. members.

I WAS 75 WHEN I RETIRED. Florence and I had a condo at a lake in Florida, Kings Point West. We would go there three or four months in the winter. Shelia called me one day. She asked me to come up for a meeting with the airport commission to discuss the renewal of the lease. I got a flight up alone that Saturday. It was very cold and icy and sleeting. I met Shelia and said, 'What a night for business.' She said, 'Well, we arranged it before we knew what the weather was going

AERO CLUB MEMBERS. FRONT ROW: OSCAR TANNENBAUM, REX TRAILER, ANN WOOD, TOM BROWN, JULIE GOLDMAN. BACK ROW: DANIEL O'CONNOR, PARKER WARD, FRANK SWEENEY, FRANK KNIGHT, CROCKER WIGHT, JOHN GRIFFIN

to be. You're here, might as well go.' It was at the hotel on Route 128, King's Grant Inn. I noticed the parking lot was packed with cars. When I opened up the door, I saw more than a hundred people I knew. The mayor and the airport commissioner and friends from Malden and all the airports were there. It was my retirement party.

The master of ceremonies said, 'Come on up, Julie, and say a few words.' I said, 'You made a mistake. You didn't bring Florence here from Florida. You couldn't do this without her here. You have to have her.' And someone tapped me on the back and it was Florence. I kissed her and gave her a few big hugs. I told stories all night and we all had a great time.

AS A KID GROWING UP, I knew a bit of Yiddish and the Hebrew alphabet. I learned the alphabet at the orphanage and I never forgot it. I was never given a bar mitzvah. Neither was my brother. But we got by alright without it.

I used to teach in a class at Harvard once a month for my Air Force unit. I was the commander of the 9221st transport squadron. One day, German was written on the board. I could figure out what it was because of my Yiddish.

I never gave anti-Semitism any thought until around 1990. I went to a meeting at Boeing in Seattle with all the fellows from the Army Air Corps. A fellow there had been one of my pilots in World War Two going over the ocean to India. He had become the head of the FAA for the United States operations. He was a skinny guy but he was pretty smart. I took him and his wife out to dinner.

He said, 'Why didn't you stay in the military. You were the chief pilot of the outfit.' I said, 'I was beyond the promotions. I didn't get any promotions and I don't know why.' He said, 'Oh, you're Jewish.' He went to St. Joseph's in Missouri in the B25 school. He said, 'There was an unwritten rule never to put a Jew in for promotion.' He told me that outright in front of Florence. I said, 'Now you see why I didn't stay.'

Something in my heart told me I shouldn't stay, I wasn't going to go anywhere yet I was doing everything that I should. The others were two grades above me all the time. Then I started to look back in my records. When I got the promotion, it was because of Joe Walker. I never thought about it until my friend and I had that conversation. Then I realized Walker had given me a superior rating deliberately because nobody did that.

When I was in Seattle with Florence, the fellows were talking. They said something about the Jews and the Israeli Air Force. I said, 'What do you mean?' I said, 'You don't talk like that in front of me.' One of the guys said, 'What are you, Jewish?' And I said yes. He shut up. Yet 99 percent of the guys were good guys.

YEARS AGO I WENT TO a family funeral, and my father said, 'Why do you work so hard?' I was working seven days a week and seven nights. I'd drive all night and fly all day. I was on the go every day of the week—no vacations. He said, 'I don't think you should work so hard.' I said, 'I'm trying to make a million dollars.'

REVERE AIRWAYS AND REVERE AVIATION OPERATED FOR ALMOST 42 YEARS BEFORE CLOSING ITS DOORS.

He said, 'Never.' I just wanted to be happy and healthy and not have to worry about money. Florence felt the same way.

MY FATHER LIVED TO BE an old man. I was called to his house when he had his heart attack. I got there and the medic had him on a stretcher. There was the ambulance outside and they weren't taking him out of the house. The medic said, 'Well, we wanted to know who was going to pay.' I blew my cork. I said, 'I'll pay, get him to the hospital.' He got to the hospital but he died while he was there. They said he tried to climb out of the stretcher. They had put the sides up. He climbed out and had another heart attack. He was stubborn. He died in May of 1966.

I was treated poorly as a kid, but I never held it against him. He was an unhappy man after he lost my mother. From then on he was bitter. I was just doing my own thing. Florence used to ask me, 'How can you visit him? He was so mean to you when you were young.' I used to take her with me when he still lived on the farm. I said, 'Well, he's still my father.' Then I'd get mad at him and wouldn't go for a while. Florence would make me go see him again. I supported him with a lot of things. He and my grandfather never thought that I'd be anything but a truck driver. And I was, for awhile.

Saul and I were very close. He did homework twice as fast as anybody in school. He had a good freehand. He'd draw a picture just sitting there, and it would look just like a photo. He was good in art but he never pursued that. Saul started drawing pictures after he came to Malden and worked with us in the trucking business. He also played the mandolin and guitar. He could play just like a top-notch guy. Even though he had rheumatic fever as a kid, he worked in the trucking business all those years. We sent him to take care of the freight in Lewiston, Maine, where he met Bea. That ended everything he did. They got married and took off for California.

Saul died in Oklahoma at his daughter's house. He was coming from a visit in Massachusetts. He was on his daughter's farm, alone, and decided to split some logs. They found him lying on the ground when the family got back from town. He was still alive. They called an ambulance to get him. On the way to the hospital, the ambulance ran out of oxygen. The rheumatic fever he got in the orphanage always bothered him. That's why they never took him in the military.

I see Lillian once in a while. She lives in Sharon, where Florence is buried. She's always on my case to take care of my health, but we get along good.

The last time I remember seeing Aunt Clara, my father's second wife, was in New York. I was walking with Florence through Times Square. Clara saw me and started to yell, 'Julie, Julie.' She was married at the time. She said, 'You know, I want to get together with your father. You can do it. You tell him what I want.' I said, 'No way, I'm not going to tell him. You tell him yourself.' She tried to talk to him. After all she had done to us. She stole all the furniture out of the house. Florence said, 'What a crazy lady.' Florence only met her once or twice.

Izzy, Paul and I bought the farm from my father for $2800. We were going to use it for a summer place. It stopped being a working farm long before that and it was just lying dormant. After we bought it, we used to drive down on weekends but Florence didn't like it. She didn't like the bugs. The town of Middleboro took it away from us because of $5000 of unpaid taxes. Oceanspray bought the house and 50 acres for several million.

Hyman, my half brother, lives down the Cape near Provincetown, and he has a big home there and a boat. He was in the Navy during the war and is Leonard Bernstein's first cousin. He's married and retired now.

My stepsister, Dolly, lived in Chicago, Washington and finally California. She had cancer, and one day when she wasn't feeling well, her husband took her to the hospital. They talked and he said that he would see her in the morning. He went home and she died during the night. Dolly was a young kid, only 59 years old.

Sonny worked for the post office in Boston as a sub

when he left the state guard. When they called him back into the military, he had worked there for four or five months. When the war was over he had priority and went to work for them again. He worked in the post office for the rest of his working life. He decided that he wanted to be in Florida. Florida was a popular place. They'd post the position plus they wanted $1500. Sonny paid the $1500 and traded his job location for one in Florida.

Uncle Iz went in the manufacturing business. He made women's dresses. After the war he and his wife opened a delicatessen and moved to Florida.

My brothers-in-law Izzy and Paul both worked in the Interstate Trucking Business. They were both in the teamsters union, a tough union. They both retired from teamsters and had a good pension fund. Izzy passed away from cancer when he was 60. His father and youngest sister Esther died from cancer.

Florence and I went to a synagogue for awhile but it was useless for me. I have love and respect for the people in Israel, but I get mad at the religious fanatics there. They create a lot of problems. On the Jewish holidays we'd have the family together and have all kinds of food. Florence was a wonderful cook. Our children and grandchildren were all given bar mitzvahs and bat mitzvahs. But I'm a heathen. My Israeli friends in the Air Force wanted to have a bar mitzvah for me at the Wailing Wall. I said, 'No use. I'm a lost cause.' I wouldn't do it. What was I trying to do, fool somebody? It's what's in my heart that's good. I'm not interested. I don't care what religion you are. If you got a good heart, you're okay.

IN 1986, MY WIFE HAD an operation for colon cancer. Everything worked out well and we never gave it another thought. Years later we went to Florence's cousin's bar mitzvah. I brought Florence little goodies from the buffet. She couldn't eat anything. We were very close to home on Lake Street in Peabody. We were driving by the cemetery when she started to be sick. I pulled into the cemetery. We got home and I called Shelia and told her we were going to the hospital.

The emergency room doctor looked at her. The doctor told us she had to be seen by someone who could look into her stomach. Dr. Aranson took x-rays. He called us into his office and told us there was cancer in the back of the stomach. It looked like it was malignant. There was no method of curing it. He recommended another doctor for surgery to bypass that area of the stomach. While he was talking to us, Florence looked at me and she started to cry. She said, 'I'm going to beat it.' I said, 'Yes, we'll beat it.'

We found there was no help. The fellow who did the operation was an expert. She had surgery at Beverly Hospital with a top surgeon. They bypassed the cancer by cutting tubes at the top of her stomach and stitched them together with the tubes at the bottom of her stomach but it didn't work. The doctor couldn't get to it. It was terrible. After the operation she was at the hospital for about a month. She went into rehabilitation for a month and then I brought her home. We took care of her at the house 24 hours a day with nurses.

I sat in a chair in my living room. Florence was in

me to the hospital and I had to stay there. The doctor told me I was going into renal failure. On Saturday they inserted a needle and a pipe on each side of my neck. They put me upstairs in the dialysis room. I was hooked up for five hours and they took 11 pounds off me that night. On Monday they put me on the machine again and they took off nine more pounds. The swelling in my legs went down and I felt better.

I had to watch a movie about where the kidneys are and how they work. I may have crushed one kidney completely years before. The kidney on one side was probably a little weak, and that's what was showing the albumin in my urine. I must of had heavy albumin when I came out of the service. A person can live on 20 percent of one kidney and be okay. One kidney was crushed and the other became weak and started to fail years ago.

I may have picked up a kidney problem from the truck accident when I went off the railroad bridge 50 years before. At the end of the war, I had a second accident at my headquarters. We had a concrete cellar with a trap door. I was talking to the fellow in charge of my G. I. bill, Danny Kline. My chief mechanic had opened the trap door. I didn't see it or hear it. When I fell, I landed across a beam which held the doors up, about two feet down. I fell like a sack and hung onto the beam. If that wasn't there, I probably would have cracked my head in the cellar. They reached down and pulled me up. One half of my body turned black and blue. I was bruised for a month after that. I never saw a doctor for it. I never realized that I may have crushed a kidney.

the far room dying of cancer and I was taking care of her. On a Friday night, I sat down and I couldn't breathe. My legs were swollen. I called Shelia and said, 'There's something wrong with me.' She came rushing over. My son-in-law, Lou, and daughter, Marlene came over and they said they would stay with Florence. Shelia took

Florence was dying of cancer and I needed dialysis three days a week. We hired two nurses to help us. While the nurses were here, Florence became very sick. Her nurse started giving her a lot of morphine. That was putting her to sleep. I said, 'Don't give her so much.' She said, 'I do what I have to do,' so I fired her. I didn't want Florence to be asleep all the time.

We had a lady taking care of the both of us, Ruth. Ruth had to help me in the shower because I had these holes in my arms and they were bandaged. I had cellophane over the bandages because I would get poisoned if I didn't handle it right. I had a cold, so I was going to sleep in the spare room. I went into our bedroom to say goodnight, I looked down and Florence looked up at me. She said, 'What are you going to do when I'm gone?' I said, 'You're not going.' I leaned over and I kissed her. I went into the other bedroom and I laid down.

I don't think I was asleep 30 minutes when Ruth woke me up. She said Florence passed away. I went back in there and I touched her. She was warm. I shook her a little bit and she was gone. The nurse was to call the doctor first before she let anyone else know. She had called him, and then she told me. I immediately called all the kids and the grandchildren and had them come over. There was a lot of crying and sadness but I wanted her family to be around her before the ambulance came to take her away. They wanted to see her one last time, and that was it. I lost her. It was 1992.

It was stomach cancer. That was one lady. If you knew her you would have loved her. Wherever I took her, whoever met her, everyone loved her. This started when we were kids. I'd take her with me everywhere. I took her around the world. Florence had motion sickness, but she went with me everywhere I went in the world, and never complained. She went to Israel with me 14 times. I had 100 trees planted in Israel in Florence's name. We had a friendship for 66 years until she passed away. We were married 63 years. She was a wonderful lady and I loved her so. I miss her so much.

I was at the cemetery a couple of months ago. I visit Florence's grave in Sharon, Massachusetts regularly. My mother is buried in Woburn, Massachusetts. The stone has to be refurbished. It has three little birds on a branch etched on to it. With eighty years of wind and rain and snow, you can barely see the little birds and branch there. I'll have to take care of that.

Index

Epilogue

Julie Goldman, 87, still flies a few hours a year but cannot fly alone as he does not have a current medical. Every two years he flies with a flight examiner who makes sure he's up to date on all the rules and flying regulations, and who then makes him current regarding commercial pilot regulations. He can then fly with anyone who has a current medical certificate and a certificate for the airplane they are in.

Catherine Andrews spent two years interviewing, transcribing, editing, and formatting Julie's book. Born in Boston, she grew up in New Hampshire and Massachusetts. She studied at Northeastern University before moving to Ohio where she earned a B. A. in English at the University of Dayton. She has worked in publishing and public relations for five years. Catherine resides in Epping, New Hampshire, with her husband, Neil, and daughter, Samantha.